Biology of Plankton

Papers by
Thomas A. Clarke, Tibor Farkas, N. J. Antia, A. F.
Carlucci, H. Fudge, L. V. Shannon, Stuart Patton,
Richard Forward, G. A. Robinson, Ruth Patrick,
A. E. Bailey-Watts, Z. Kalnina, Wolfgang H. Berger,
E. D. S. Corner, J. S. Kittredge, Charles F. Wurster,
Jr., James L. Cox, Alan D. Hecht, David W. Menzel,
John H. Ryther, George W. Ware, Robert C. Harriss,
Peter Richerson, et al.

MSS Information Corporation
655 Madison Avenue, New York, N.Y. 10021

724345

Library of Congress Cataloging in Publication Data
Main entry under title:

Biology of plankton.

Includes bibliographies.
1. Plankton--Addresses, essays, lectures.
I. Clarke, Thomas A.
QH91.8.P5B53   574.92   72-7007
ISBN 0-8422-7016-7

TABLE OF CONTENTS

Ecological Studies during Project Sealab II . . . Clarke, Flechsig and Grigg   7

The Effect of Environmental Temperature
on the Fatty Acid Compositiion of Crustacean Plankton . . . . . Farkas
Herodek   25

Enolase Activity in Marine Planktonic Algae. . Antia, Kalmakoff and Watt   36

Bioassay of Seawater. I. A ^{14}C Uptake
Method for the Determination of Concentrations
of Vitamin B12 in Seawater . . . . Carlucci and Silbernage   42

Biochemical Analysis of Preserved Zooplankton . . . . Fudge   51

Polonium-210 and Lead-210 in the Marine Environment . . . . . . Shannon,
Cherry and Orren   53

Food Value of Red Tide *(Gonyaulax polyedra)* . . . . Patton,
Chandler, Kalan, Loeblich, Fuller and Benson   64

Red and Far-Red Light Effects on a Short-Term
Behavioral Response of a Dinoflagellate . . . . . Forward and Davenport   68

Distribution of *Gonyaulax tamarensis* Lebour in
the Western North Sea in April, May and June 1968 . . . . . . Robinson   71

The Effect of Invasion Rate, Species Pool, and
Size of Area on the Structure of the Diatom Community . . . . . Patrick   75

Freshwater Primary Production by
a Blue-Green Alga of Bacterial Size. Bailey-Watts, Bindloss and Belcher   83

Strontium-90 Concentration Factors of
Lake Plankton, Macrophytes, and Substrates . . . Kalnina and Polikarpov   88

Planktonic Foraminifera: Field Experiment
on Production Rate . . . . Berger and Soutar   93

Biochemical Studies on the
Production of Marine Zooplankton . . . . Corner and Cowey   98

Aminophosphonic Acids: Biosynthesis by
Marine Phytoplankton . . . . Kittredge, Horiguchi and Williams   132

DDT Reduces Photosynthesis by Marine Phytoplankton . . . . . . Wurster   137

DDT Residues in Marine Phytoplankton:
Increase from 1955 to 1969 . . . . Cox   142

Oxygen-18 Studies of Recent Planktonic
Foraminifera: Comparisons of Phenotypes
and of Test Parts. . . . . Hecht and Savin   146

Marine Phytoplankton Vary in Their
Response to Chlorinated Hydrocarbons . . . . Menzel,
Anderson and Randtke   151

Is the World's Oxygen Supply Threatened? . . . . Ryther   155

Interaction of Pesticides with
Aquatic Microorganisms and Plankton . . . . Ware and Roan   157

Mercury Compounds Reduce Photosynthesis
by Plankton . . . . Harriss, White and Macfarlane   187

Contemporaneous Disequilibrium, a New Hypothesis to Explain
the "Paradox of the Plankton" . . . Richerson, Armstrong and Goldman   192

CREDITS & ACKNOWLEDGEMENTS

Antia, N. J.; J. Kalmakoff; and A. Watt, "Enolase Activity in Marine Planktonic Algae," *Canadian Journal of Biochemistry* 1966, 44:449 — 454.

Bailey-Watts, A. E.; M. E. Bindloss; and J. H. Belcher, "Freshwater Primary Production by a Blue-Green Alga of Bacterial Size," *Nature,* 1968, 220:1344 — 1345

Berger, Wolfgang H.; and Andrew Soutar, "Planktonic Foraminifera: Field Experiment on Production Rate," *Science,* 1967, 156:1495 — 1497. Copyright 1967 by the American Association for the Advancement of Science.

Carlucci, A. F.; and S. B. Silbernagel, "Bioassay of Seawater. I. A ^{14}C Uptake Method for the Determination of Concentrations of Vitamin B12 in Seawater, " *Canadian Journal of Microbiology,* 1966, 12: 175 — 183.

Clarke, Thomas A.; Arthur O. Flechsig; and Richard W. Grigg, "Ecological Studies during Project Sealab II," *Science,* 1967, 157:1381 — 1389. Copyright 1967 by the American Association for the Advancement of Science.

Corner, E. D. S.; and C. B. Cowey, "Biochemical Studies on the Production of Marine Zooplankton," *Biological Reviews,* 1968, 43:393 — 426.

Cox, James L., "DDT Residues in Marine Phytoplankton: Increase from 1955 to 1969," *Science,* 1970, 170:71 — 73. Copyright 1970 by the American Association for the Advancement of Science.

Farkas, Tibor; and Sandor Herodek, "The Effect of Environmental Temperature on the Fatty Acid Composition of Crustacean Plankton," *Journal of Lipid Research,* 1964, 5:369 — 373.

Forward, Richard; and Demorest Davenport, "Red and Far-Red Light Effects on a Short-Term Behavioral Response of a Dinoflagellate," *Science,* 1968, 161:1028 — 1029. Copyright 1968 by the American Association for the Advancement of Science.

Fudge, H., "Biochemical Analysis of Preserved Zooplankton, " *Nature,* 1968, 219:380 — 381

Harriss, Robert C.; David B. White; and Robert B. Macfarlane, "Mercury Compounds Reduce Photosynthesis by Plankton," *Science,* 1970, 170:736 — 737. Copyright 1970 by the American Association for the Advancement of Science.

Hecht, Alan D.; and Samuel M. Savin, "Oxygen-18 Studies of Recent Planktonic Foraminifera: Comparisons of Phenotypes and of Test Parts," *Science,* 1970, 170:69—71. Copyright 1970 by the American Association for the Advancement of Science.

Kalnina, Z.; and G. Polikarpov, "Strontium-90 Concentration Factors of Lake Plankton, Macrophytes, and Substrates," *Science,* 1969, 164: 1517—1519. Copyright 1969 by the American Association for the Advancement of Science.

Kittredge, J. S.; M. Horiguchi; and P. M. Williams, "Aminophosphonic Acids: Biosynthesis by Marine Phytoplankton," *Comparative Biochemistry and Physiology,* 1969, 29:859—863.

Menzel, David W.; Judith Anderson; and Ann Randtke, "Marine Phytoplankton Vary in Their Response to Chlorinated Hydrocarbons," *Science,* 1970, 167:1724—1726. Copyright 1970 by the American Association for the Advancement of Science.

Patrick, Ruth, "The Effect of Invasion Rate, Species Pool, and Size of Area on the Structure of the Diatom Community," *Proceedings of the National Academy of Sciences,* 1967, 58:1335—1342.

Patton, Stuart; P. T. Chandler; E. B. Kalan; A. R. Loeblich III; G. Fuller; and A. A. Benson, "Food Value of Red Tide *(Gonyaulax polyedra), Science,* 1967, 158:789—790. Copyright 1967 by the American Association for the Advancement of Science.

Richerson, Peter; Richard Armstrong; and Charles R. Goldman, "Contemporaneous Disequilibrium, a New Hypothesis to Explain the 'Paradox of the Plankton' ", *Proceedings of the National Academy of Sciences,* 1970, 67:1710—1714.

Robinson, G. A., "Distribution of *Gonyaulax tamarensis* Lebour in the Western North Sea in April, May and June 1968," *Nature,* 1968, 220:22—23.

Ryther, John H., "Is the World's Oxygen Supply Threatened?" *Nature,* 1970, 227:374—375.

Shannon, L. V.; R. D. Cherry; and M. J. Orren, "Polonium-210 and Lead-210 in the Marine Environment," *Geochimica et Cosmochimica Acta,* 1970, 34:701—711.

Ware, George W.; and Clifford C. Roan, "Interaction of Pesticides with Aquatic Microorganisms and Plankton," *Residue Review,* 1970, 33:15—45.

Wurster, Charles F. Jr., "DDT Reduces Photosynthesis by Marine Phytoplankton," *Science,* 1968, 159:1474—1475. Copyright 1968 by the American Association for the Advancement of Science. Robinson, G. A., 81

Ecological Studies during Project Sealab II

Thomas A. Clarke, Arthur O. Flechsig, Richard W. Grigg

During August, September, and October, 1965, the U.S. Navy's Special Projects Office and Office of Naval Research conducted Project Sealab II off La Jolla, California. The main purpose of the project was to evaluate the performance of men and equipment in a high-pressure, underwater environment (*1*). *Sealab II*, an underwater habitat, was placed on the bottom for 45 days. Three ten-man teams lived in *Sealab II* for about 2 weeks each. The men lived at ambient pressure for the entire period and had access to the surrounding water through an open entryway in the bottom of *Sealab II*.

We participated as divers, one of us on each team. Thus our observations cover the entire period during which *Sealab II* was on the bottom. During this time we studied the ecology of the sand bottom around the site and observed on a day-by-day basis the organisms attracted to the site. We recorded abundances, behavior, and food habits. Most of our observations were of areas adjacent to *Sealab II*, but we were also able to make several surveys of the sand bottom at locations well removed from the site. Thus we are able to compare the fauna attracted to *Sealab II* with the normal sand-bottom community. We believe this was the first opportunity biologists have had to conduct a continuous underwater study of marine organisms.

Site and Environment

Sealab II was placed approximately 1400 meters from shore at a depth of 61 meters in a small, gently sloping valley (slope of 10 degrees) near the main axis of Scripps Submarine Canyon (Fig. 1). The bottom sediments at the site were silty-sand, and the bottom was essentially featureless, with only minor hummocks, 1 to 2 centimeters in vertical relief. Thirty-five meters northwest of *Sealab II*, the bottom steepened, and slightly beyond, at a depth of 76 meters, a vertical rock cliff dropped into a tributary of the main canyon. Other than the canyon walls, the nearest rocky bottom was 1500 meters away, near shore.

The largest object on the bottom was *Sealab II*, a cylinder 17.5 meters long, 3.7 meters in diameter, and 9.1 meters high at the central conning tower. The entire structure was painted white. There were 11 circular viewing ports 60 centimeters in diameter. These viewing ports permitted more or less constant monitoring of events occurring outside. Six of them were equipped with external 1000-watt incandescent lamps with reflectors. Various combinations of these external lights were on during the project. Less intense light from inside was visible through all the viewing ports.

Three other large objects were placed on the bottom: the personnel transfer chamber, used to transport men to the surface under pressure; the benthic lab, which housed a communications transformer; and a power transformer. All were cylinders about 4 meters high and 2 meters in diameter and were colored orange. Only the transfer chamber was lighted. Slightly smaller objects in the assemblage included a wire-mesh fish cage, three underwater instrument stations, and an emergency breathing chamber.

Sealab II was lowered to the bottom on 26 August 1965. The following day the personnel transfer chamber and power transformer were lowered, and *Sealab II* was inspected by divers.

On 28 August, day 1, the first team of divers descended to *Sealab II*. Some of the viewing port covers were not removed until day 2, and few outside observations were made before day 3.

The first team returned to the surface on 12 September. The second team was down from 12 September to 26 September; the third, from 26 September to 10 October, day 44. *Sealab II* and the other large structures were raised the following day. A brief dive was made to inspect the site 19 days later.

During the project the water temperature on the bottom ranged between 10° and 13°C. Visibility was often poor near the bottom, owing to suspended detritus and sediment stirred up by the activity of divers; it ranged from 0 to 10 meters. Wave surge was rarely noted, but it once reached a horizontal displacement of 20 centimeters. Speeds of persistent currents ranged from 0 to 30 centimeters per second but were usually less than 10 centimeters per second.

Methods

Since the atmosphere in the living quarters was at ambient pressure, divers were able to work in the water for long periods at depths between 50 and 90 meters without having to undergo decompression. Individual divers averaged about 1 hour per day in the water. Most diving activity occurred during the daylight hours and within 10 meters of *Sealab II*.

The smaller bottom organisms were sampled with a hand corer that took a 35-square-centimeter sediment sample to a depth of 5 centimeters. We estimated the densities of larger organisms from counts made along transects by means of a fish rake (2), and also from counts made within 1-meter-square quadrats. The locations of the transects and core samples are shown in Fig. 1. Zooplankton were sampled by divers with hand nets or push nets. Estimates of total abundances for mid-water fishes were obtained by extrapolation from counts for a representative volume. Fish were collected by spear for identification and stomach-content analysis, but none were speared until the final 2 weeks of the project.

In addition to the zooplankton and smaller benthic invertebrates, 11 species of larger invertebrates, 43 species of fishes, and one species of mammal (sea lion) were observed during the project. Of these, ten invertebrates and 17 fishes appeared to be normal inhabitants of the sand bottom, while two invertebrates, 17 fishes, and the sea lion were attracted to and associated with *Sealab II*. The remaining species of fishes were observed so rarely that they cannot be assigned to either group (3).

The Sand-Bottom Community

Although the site area was generally typical of local sandy bottoms at these depths in terms of sediment, relief, and currents, the proximity of the canyon may have influenced the types and numbers of organisms present, especially as many sand-bottom organisms are locally more abundant along the canyon edges than on the open sand.

The most abundant organisms in the core samples were polychaetes, especially errant species (Table 1). Next in order of abundance were nematodes, foraminifera, crustacea, and ophiuroids. Seaweed debris was found in almost all cores. The commonest large invertebrates (Table 2) were the sea stars *Petalaster foliata* and *Astropecten verrilli*, the sea urchin *Lytechinus pictus*, and the sea pen *Stylatula elongata*. A small octopus was also a member of the sand-bottom community, but it aggregated in large numbers on *Sealab II* and other structures and, therefore, our observations do not reflect its normal density. Except for the octopus, the normal distribution and abundances of bottom invertebrates were apparently unaffected by the presence of *Sealab II*. Push-net samples of zooplankton

taken just above the bottom contained mostly mysids and amphipods. The average volume of the plankton obtained in four such tows was 0.4 cubic centimeter per cubic meter of water sampled.

Table 1. Data for small invertebrates from the core samples. Only subgroups that appeared in more than 60 percent of the cores are included; those present in over 90 percent of the cores are starred. Density is based on combined data from 28 cores. Biomass is given for the major groups only.

Group	Density (No./m²)	Wet wt. (g/m²)
Seaweed debris*		
Foraminifera*	6,660	
Nemertea	316	
Nematoda*	7,920	0.374
Polychaeta (total)*	10,360	4.660
Errantia (total)*	8,200	
Nephthydidae*	6,600	
Phyllodocidae	656	
Glyceridae	184	
Sedentaria (total)*	2,160	
Spionidae	1,288	
Cirratulidae	449	
Mollusca (total)	1,449	
Gastropoda	459	
Pelecypoda	990	
Crustacea (total)*	3,840	5.950
Ostracoda	530	
Copepoda	1,202	
Amphipoda*	1,162	
Decapoda	612	
Others	334	
Ophiuroidea*	1,350	57.000

Small flatfishes were the most abundant fishes on the open sand. The common species, in order of decreasing abundance, were the longfin sand dab, *Citharichthys xanthostigma;* the hornyhead turbot, *Pleuronichthys verticalis;* the Pacific sand dab, *C. sordidus;* and the California tongue fish, *Symphurus atricauda*. The densities recorded (Table

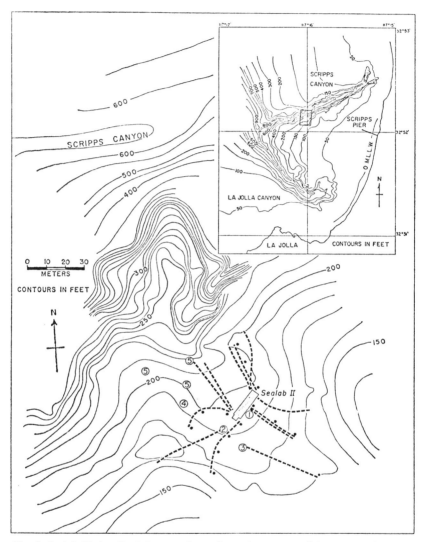

Fig. 1. *Sealab II* site. Shaded rectangle on inset shows area of large map. 1, Power transformer; 2, personnel transfer chamber; 3, benthic laboratory; 4, fish cage; 5, instrument stations. Dashed lines and black dots show locations of fish-rake transects and paired core samples, respectively.

10

2) are similar to those recorded for flatfishes in shallower water (4). The fishes' stomachs contained mostly small benthic organisms and, to a lesser extent, hypoplanktonic mysids and amphipods. The bluespot goby, *Coryphopterus nicholsi*, was occasionally observed on the open sand in densities up to 50 per square meter. We do not, however, have a reliable estimate of their normal density, either because their distribution was extremely patchy or because they were often overlooked on account of their small size. In addition to these bottom species, schools (5) of five to 25 pink sea perch, *Zalembius rosaceus*, were seen just above the bottom. Individual fish would occasionally dip down and peck at the sand. This pecking left behind a characteristic pattern of pockmarks on the sand that was often seen even when the fish were not. Owing to the patchy distribution and wariness of the fish, we were unable to reliably estimate their density. All of the fishes mentioned above were apparently repelled from the *Sealab II* site. They were seen only over the open sand and never within 10 meters of *Sealab II* (Fig. 2).

Carnivores that preyed on larger organisms were rare on the open sand. Pacific angel sharks, *Squatina californica*, were seen most often (Table 2). Several California halibut, *Paralichthys californicus*, were seen momentarily near *Sealab II* but were not seen on the open sand, probably because of their extreme wariness. The California lizard fish, *Synodus lucioceps*, was seen occasionally both on the open sand and near *Sealab II*.

California scorpion fish, *Scorpaena guttata*, and calico rockfish, *Sebastodes dalli*, two species that usually inhabit rocky bottoms, were also found on the sand but were usually associated with mats of seaweed debris. A mat approximately 4 square meters in area would

often harbor about six of each species. Scorpion fish were attracted to *Sealab II*, but the calico rockfish were neither repelled from, nor particularly attracted to, the site. We do not have precise estimates of the normal density for either species on the sand bottom, but they were at least as numerous as all other higher carnivores of the sand bottom.

Table 2. Data for commonly occurring larger invertebrates and fishes of the open sand bottom (16). Size is in terms of total length or maximum radius. Density is based on combined data from all transects. Biomass was estimated by weighing preserved fish of modal size.

Species	Length or radius (cm)	Density (No./ 100 m²)	Biomass (g/ 100 m²)
Sea pen			
Stylatula elongata	10–15	5.0	
Sea stars			
Astropecten verrilli	3–5	6.6	
Petalaster foliata	8–10	2.8	
Sea urchin			
Lytechinus pictus	1–3	5.0	
Octopus	10		
Pacific angel shark	60–100	1.3	3794
Flatfishes			
Pacific sand dab	14–20	3.2	182
Longfin sand dab	12–20	5.4	254
Hornyhead turbot	9–15	3.8	133
California			
tonguefish	5–10	1.0	6

There were three general trophic levels in the sand-bottom community (Fig. 3). Most of the invertebrates probably fed on seaweed debris on or buried in the sand. The primary carnivores, largely small fishes, fed mostly on the benthic invertebrates. The standing crop of small benthic and hypoplanktonic invertebrates was about 70 grams per

11

square meter. The standing crop of those primary carnivores for which we have reliable estimates was 5.75 grams per square meter. The ratio of standing crops suggests that the primary carnivores could be supported primarily by resident food.

The higher-order carnivores had a much greater biomass than the primary carnivores. The estimated density of angel sharks, the only large predator sampled reliably, was only one individual per 100 square meters, but their biomass was 38 grams per square meter—almost seven times that of the primary carnivores. Inclusion of the other higher carnivores would at least double this biomass. This high biomass suggests that the higher carnivores depended on a nonresident food supply. It is possible that these predators were attracted to the area by *Sealab II* and fed on the other fishes attracted there, but, with the exception of the scorpion fish and a few lizard fish, they were seen only momentarily near the site and were not observed feeding. A great deal of previous field experience indicates that such predators are normally more abundant near the canyon edges and that their prey are primarily fishes from mid-water or the canyon.

The Attracted Fauna

Organisms were immediately attracted to *Sealab II* when it was lowered. On 27 August, the day before *Sealab II* was occupied, cabezon (*Scorpaenichthys marmoratus*) and schools of northern anchovy (*Engraulis mordax*) were seen at the site by surface divers. By day 1, blacksmith (*Chromis punctipinnis*) had taken up positions around *Sealab II*, and shiner perch (*Cymatogaster aggregata*), sharpnose sea perch (*Phanerodon atripes*), and white croaker (*Genyonemus lineatus*) were concentrated at the lights. Small octopus were hiding in many sheltered corners on

Sealab II.

That night a swarm of zooplankton and a brown rockfish, *Sebastodes auriculatus*, appeared at the illuminated entryway. On day 2, all the viewing-port covers were removed and vermilion rockfish (*Sebastodes miniatus*) and California scorpion fish (*Scorpaena guttata*) were first seen; the scorpion fish were already abundant (approximately one per square meter) over a wide area around the site but were not yet concentrated close to *Sealab II*. On day 3, sand bass (*Paralabrax nebulifer*) and rubberlip sea perch (*Rhacochilus toxotes*) were seen during the day, and schools of squid (*Loligo opalescens*), jack mackerel (*Trachurus symmetricus*), and California bonito (*Sarda lineolata*) were seen from the ports at night. California sea lions (*Zalophus californicus*) appeared midway in the project. Other species (6) appeared later, but all of

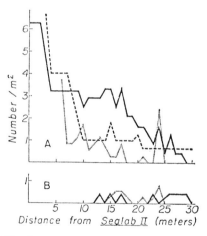

Fig. 2. Distribution and density of two bottom fishes. A, Scorpion fish, the most abundant attracted species; B, longfin sand dab, the most abundant sand-bottom fish. (Dotted line) Data for days 1–15; (dashed line) days 16–19; (solid line) days 30–43.

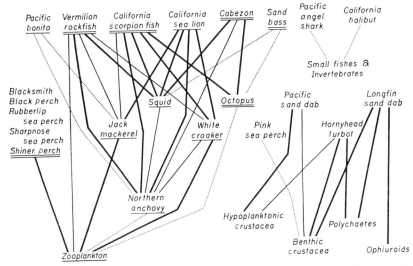

Fig. 3. Food web of the sand-bottom community and the attracted fauna. Heavy solid lines indicate predominant links. All solid lines are based on direct observation or analysis of stomach contents; dotted lines are inferred links. Attracted species are underlined—visitors singly, residents doubly.

the common species were present in large numbers by day 3.

The attracted organisms were of two types, species resident at the site and nightly visitors. The common resident species—octopus, sand bass, the embiotocid perches, scorpion fish, vermilion rockfish, and cabezon—were to be found, both day and night, either on or within a few meters of Sealab II. With the exception of the octopus, these are species that normally inhabit rocky bottoms, although three of the species— sand bass, scorpion fish, and cabezon— also wander over open sand and were seen there occasionally during the project. None of the other resident species

were ever observed over the open sand. Residents as a group increased steadily throughout the project (Fig. 4). By day 43, over 5800 individuals were present over a 500-square-meter area around the Sealab II site, a density of 11.6 individuals per square meter or 2.86 kilograms per square meter (Table 3). This biomass is 35 times greater than our maximum estimate for the biomass of fishes in the sand-bottom community. Nineteen days after Sealab II was raised, only four scorpion fish, two calico rockfish, and one rubberlip sea perch were observed at the site over a 250-square-meter area.

The visiting species appeared at the

13

Sealab II lights in the evening as natural illumination diminished and were rarely seen during the day, even though lights were on at all times. The common visitors were zooplankton, squid, anchovy, jack mackerel, bonito, white croaker, and sea lions. Except for the croaker, these are pelagic animals not normally associated with the bottom. Because the visitors appeared irregularly in schools or swarms, our estimates may not be accurate, but for most visiting species the numbers attracted at night appeared to be nearly constant throughout the project. The white croakers were an exception; their numbers increased steadily until midway of the project, when nearly all disappeared, apparently frightened away by sea lions.

The organisms attracted to the site fed almost exclusively on visiting prey (Fig. 3). Because of their greater numbers, the residents consumed more than the visitors did. Of the residents, large predators made up 61 percent of the numbers and 92 percent of the biomass. Their principal food was anchovy. The remaining residents ate zooplankton. Of the visitors, most ate zooplankton, but squid, bonito, and sea lions ate visiting fish.

Observations on Common Animals

In this section we report some of our observations on behavior and species interactions. Such data can be obtained with relative ease from an underwater installation such as *Sealab II*. Where these observations are compared with "normal" habits, our sources are Limbaugh (7) and our own observations during several years of diving experience. Information about only the more abundant species is included.

Zooplankton. Zooplankton swarmed around the lights every night. It consisted mostly of forms attracted from mid-water; there were few hypoplanktonic forms from the sand bottom. Euphausids, copepods, and zoeae formed separate dense swarms around the external lights at various times. All were attracted to the lighted viewing ports. The copepods and zoeae simply aggregated against the glass, but the euphausids formed a definite pattern, with a swarm at the bottom and two "tongues" extending up the perimeter. Those at the perimeter were swimming upward and appeared to be creating a downward current along the edges.

A 0.5-cubic-meter sample taken at night from a dense swarm outside a lighted viewing port contained approximately 30,000 euphausids (*Nyctiphanes simplex*), 8000 copepods (principally *Calanus helgolandicus* and *Paracalanus* sp.), and smaller numbers of other plankton. The volume of zooplankton in the swarm was 528 cubic centimeters

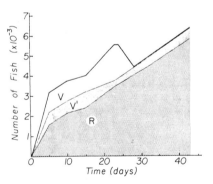

Fig. 4. Total numbers of fishes over a 500-square-meter area around *Sealab II* at various times during the project. Shaded areas show proportion of (*R*) residents, (*V*) white croaker, and (*V'*) other visitors.

Table 3. Commonly occurring organisms attracted to *Sealab II* and estimates of their abundance over a 500-square-meter area around the site at various times during the project. Biomass was estimated by weighing preserved specimens of modal size.

Species	Abundance (No./500 m²)						Biomass, day 43 (kg/500 m²)
	Day 5	Day 10	Day 15	Day 23	Day 33	Day 43	
Residents at the site day and night							
Octopus							?
Shiner perch	400	400	400	1000	1200	1650	54.45
Sharpnose sea perch	150	150	150	100	250	500	34.75
Rubberlip sea perch	50	75	75	10	50	50	16.40
Blacksmith	20	20	20	15	25	50	9.80
California scorpion fish	500	750	1000	1600	2300	2800	1092.00
Vermilion rockfish	500	750	750	750	750	750	216.75
Cabezon	10	10	10	5	5	5	7.69
Totals (fishes)	1630	2155	2405	3480	4580	5805	1431.84
Night time visitors							
Zooplankton							3.00
Squid							?
Northern anchovy	200	200	200	200	300	300	6.15
Pacific jack mackerel	400	400	400	100	200	300	30.60
White croaker	1000	1000	1000	1800	10	10	271.80*
Totals (fishes)	1600	1600	1600	2100	510	610	36.75

* Croaker biomass is for day 23 and is not included in total.

per cubic meter. The volume of samples collected over the nearby sand averaged 0.4 cubic centimeter per cubic meter. Thus the abundance of zooplankton around the lights was greater by about three orders of magnitude than that over the open sand. On the basis of only those volumes where zooplankton were densely aggregated, the standing crop at night around *Sealab II* was estimated to be 3 kilograms.

White croaker, jack mackerel, sharpnose perch, shiner perch, and blacksmith ate zooplankton almost exclusively. The perch and blacksmith fed only at the periphery of the dense swarms and not in their midst. Dense swarms were never seen around the lights when croaker and jack mackerel were feeding, but the movement of the fish was probably as great a factor in disrupting swarms as their feeding was.

Octopus. The small octopus of the species observed normally inhabits the sandy bottom. It burrows in the sand or hides under or in solid objects such as shells, bottles, and cans. Large numbers were attracted to *Sealab II* immediately. As many as six were found hiding beneath a single port cover on day 2. Each time the personnel transfer chamber was raised, approximately 40 octopi were still clinging to it when it reached the surface. Because of the secretive behavior of the octopus, no rigorous estimates of abundance were possible, but several hundred individuals were probably hiding on *Sealab II* and other objects at the site. Their rapid accumulation indicates that the species observed is more mobile than rock-bottom species of octopus, which appear to have restricted home ranges.

Initially the small octopus was seen only when equipment was moved or

Fig. 5. White croaker, *Genyonemus lineatus*, feeding on zooplankton at a viewing port. [Official U.S. Navy photograph by J. D. Skidmore]

used. Later in the project, as many as 12 at one time were seen at night, suspended in the water in front of the lighted viewing ports, apparently feeding on zooplankton which they caught with outstretched tentacles. We did not observe any predation on octopus, but octopus remains were found in the

16

stomachs of scorpion fish and cabezon. *Squid*. Squid (*Loligo opalescens*) were first observed on day 3 and appeared irregularly, but with increasing frequency and numbers, from then on. They were seen only at night; schools of three to five and later up to 20 squid would dart into the lighted area and gradually disperse. Squid fed on anchovy and were in turn eaten by vermilion rockfish, scorpion fish, cabezon, and sea lions. Their common occur-

rence in predator stomachs indicates that they were present more often than they were observed.

Northern anchovy. Anchovy are normally pelagic, schooling fish. Northern anchovy (*Engraulis mordax*) were observed at the site by surface divers on 27 August and were seen from *Sealab II* on day 2. They were seen at the viewing ports most nights thereafter, appearing in small schools at irregular intervals. They schooled according to

Fig. 6. A vermilion rockfish, *Sebastodes miniatus*, holding a northern anchovy, *Engraulis mordax*, in its mouth.

size; two size classes (length) were seen: 12 centimeters and, less often, 7 centimeters. Most anchovy were visible only at the edge of the light. Those that entered the well-lighted zone darted blindly back and forth across the port for a few seconds and then vanished quickly or were eaten. Individuals observed in the lighted volume were regularly preyed upon by scorpion fish, vermilion rockfish, and sea lions, and less often by squid and croaker. Anchovy were not seen feeding.

Sometimes, when grasped by a predator, an anchovy would escape and leave the predator disgorging scales. This means of escape was evidently possible because anchovy have deciduous scales; it may be analogous to the escape of lepidopterans from spider webs (8). Such a mechanism would have an obvious selective advantage for fishes subject to heavy predation and may, therefore, explain why deciduous scales are characteristic of a number of small, pelagic, schooling fishes.

Jack mackerel. Jack mackerel (*Trachurus symmetricus*) are common, coastal-pelagic, schooling fish. They were first seen at *Sealab II* on day 3. Schools of 150 to 200 individuals appeared most nights thereafter; they were rarely seen during the day. In some schools as many as one third of the individuals had wounds or scars. Jack mackerel approached the external lights as a tight school but dispersed to feed on individual zooplankters. Individuals assumed all orientations while feeding. Jack mackerel were eaten by scorpion fish, vermilion rockfish, bonito, and sea lions. They reacted differently to different predators. When they were attacked by scorpion fish or rockfish, only the individual actually attacked reacted. In contrast, when they were attacked by bonito or a sea lion, all would school up and dash off as a unit. Sea lions and bonito commonly prey on jack mackerel, while scorpion fish and vermilion

rockfish do not. Furthermore, sea lions and schools of bonito attacked with prolonged, high-speed approaches, most often from above. Scorpion fish and rockfish attacked as individuals in short, sudden lunges, usually from below. The alarm response could, therefore, have been determined either by recognition of the particular species as a predator or by the type of attack.

White croaker. The white croaker (*Genyonemus lineatus*) is a schooling species commonly found over sand bottoms in shallow water. Schools of them were seen on day 1, and their numbers increased until day 24, when sea lions began feeding at *Sealab II*. Few were seen during the day, but at night individuals milled around the lights and at times pressed so thickly against the viewing ports that they blocked the view completely (Fig. 5). They fed nonselectively on zooplankton. At viewing ports with strong interior lights, the croakers pressed their heads against the glass and inhaled the swarms of zooplankton.

About 1 percent of the croakers were either parasitized by copepods or had tumors around the mouth and gills. Some bore wounds or scars. Vermilion rockfish and, more often, scorpion fish preyed on croakers, but the croakers seemed little disturbed by these predators. They were, however, often alarmed by the sudden appearance of a school of jack mackerel. Also, if jack mackerel were alarmed and moved off as a school, the croakers that were mixed in with them would join the school and move off, but only for a short distance. Croakers feeding at the viewing port a meter away from schools were unperturbed by these alarms.

After sea lions began diving, on day 24, the croakers were seen only in tight, columnar schools. After 4 nights of sea-lion attacks the number of croakers appearing at the lights was less than 1 percent of the former number. The sea lions did feed on croakers, but each of them would have had to consume over 200 fish per night to account for this disappearance. It is more likely that the croakers were frightened away. Their disappearance was responsible for the drop in total estimated number of fishes between days 24 and 28 (Fig. 4).

Sharpnose sea perch. Sharpnose sea perch (*Phanerodon atripes*) were noted at *Sealab II* on day 1, and they increased in abundance slowly throughout the project. They aggregated but did not school. During the day individuals swam in mid-water above and around *Sealab II*. At night they were concentrated around the external lights and viewing ports, where they swam in and out of the lighted volume. Once a sharpnose sea perch was seen cleaning a vermilion rockfish in mid-water.

They did not feed on dense plankton swarms but, rather, selected individual zooplankters. One stomach examined contained mostly one type of crab zoea. The sea perch would occasionally compete with each other or with blacksmith for the same zooplankter.

Shiner perch. Shiner perch (*Cymatogaster aggregata*) inhabit bays or deeper water on the open coast. They were noted at *Sealab II* on day 1 and increased in abundance throughout the project. During the day large aggregations hung above *Sealab II*. Smaller tight schools were also seen around *Sealab II* and the personnel transfer chamber, and at unlighted objects such as the fish cage and an instrument station. At night, shiner perch aggregated around the ports and fed on individual zooplankters, much like the sharpnose sea perch. Shiner perch near the light swam on their sides or even upside-down, keeping their dorsal surfaces toward the light.

Rubberlip sea perch. Rubberlip sea perch (*Rhacochilus toxotes*) normally inhabit rocky bottoms. They appeared at *Sealab II* on day 3; from day 5 on, 50 to 75 individuals were usually present. Occasionally small schools or individuals would pick at algal detritus on the sand. At night individuals occasionally moved into the lights and fed on zooplankton, but most remained at the edge of the field of view. The stomach from a fish collected in the daytime contained an alpheid shrimp from the bottom and many euphausids and copepods from mid-water.

Blacksmith. Blacksmith (*Chromis punctipinnis*) are rocky-bottom fish and are most abundant in shallower water. They were noted at the site on day 1; by day 5, 20 adults were closely associated with *Sealab II*. Blacksmith were repelled by flashlights and were more numerous at the dimly lighted ports. They fed on individual zooplankters; their stomachs contained euphausids, copepods, and crab zoeae.

In their usual habitat blacksmith school and feed on zooplankton in mid-water by day; at night they retire to crevices in the rocks. In contrast, blacksmith at *Sealab II* did not school and did not move up into mid-water but fed around the ports both day and night. They appeared to have a preferred location which they defended against sharpnose and rubberlip sea perch. They occasionally nipped at each other while competing for food.

Vermilion rockfish. Vermilion rockfish (*Sebastodes miniatus*) are known to be abundant in the canyon and over rock outcrops and cobble patches in deep water. A group of 150 individuals was observed at the edge of the canyon 35 meters away from the *Sealab II* site. Vermilion rockfish were among the first arrivals at the site. By day 10, as many as 750 were present; thereafter their number remained nearly constant. Changes in size frequency, however, suggested that interchange with the canyon aggregations occurred. Most of the rockfish hung in mid-water as a

Fig. 7. California scorpion fish, *Scorpaena guttata*, piled in the entry-way of *Sealab II*. Several divers were stung while working in this area. The fish above the bottom are shiner perch, *Cymatogaster aggregata*.

loose aggregation around and above *Sealab II*. Individuals swam around the ports, the external lights, and the entryway, and occasionally rested on *Sealab II*. When the sea lions first appeared, the vermilion rockfish gathered in sheltered areas on *Sealab II*, but within 3 days they had resumed their normal behavior. They were curious and often followed divers. A tapping or rubbing on the port glass from inside would attract several to the port temporarily.

The vermilion rockfish fed all night and were observed eating anchovy, jack mackerel, squid, individual zooplankters, and occasionally croaker. Individuals with swollen guts did not cease feeding; one held an anchovy in its mouth for 2½ hours before it could

swallow it (Fig 6). The stomachs of three of five rockfish collected during an afternoon contained fresh anchovies and squid, indicating that they fed during the day as well.

California scorpion fish. Scorpion fish (*Scorpaena guttata*) usually live as solitary individuals on rocky substrate. They also move across open sand and aggregate in kelp debris. Large aggregations are known to occur suddenly during the spring and summer months around both natural and artificial reefs (9). At *Sealab II* they were first seen on day 2 and increased in abundance throughout the project. At the end they were the most abundant fish; it was estimated that 2800 individuals were present on the bottom within a 500-

square-meter area.

Scorpion fish were strongly thigmotactic and were almost always in contact with solid objects, irregularities of the bottom, or other scorpion fish. They were most abundant in the best-lighted areas, where prey were also most abundant. They literally piled up in and near the entryway of *Sealab II* (Fig. 7), and, because of their poisonous spines, were a hazard to divers. Densities up to 27 individuals per square meter were recorded in this area. Scorpion fish in densities up to 5 per square meter were also noted around smaller unlighted objects such as the instrument stations and fish cage. Figure 2 shows the decline in abundance of scorpion fish with distance from *Sealab II*.

Fresh octopus and anchovy were found in the stomachs of specimens speared during the day. At night, scorpion fish were observed feeding on anchovy, white croaker, jack mackerel, and squid, which they captured by sudden lunges from resting places on *Sealab II* or the bottom. Scorpion fish were not preyed upon.

Cabezon. Cabezon (*Scorpaenichthys marmoratus*) usually inhabit rocky bottoms but are occasionally seen on the open sand. One individual was observed on the sand near *Sealab II*, and two were seen in the nearby canyon. From five to ten cabezon were situated on horizontal surfaces on *Sealab II* throughout the project, each individual frequenting a particular spot or crevice. At night, cabezon fed on anchovy and squid, which they captured by lunging from their resting places. Stomachs of cabezon collected in the daytime contained fish remains and cephalopod beaks, and, in two cases, fresh squid and octopus.

California sea lion. Sea lions (*Zalophus californicus*) are common locally and were observed playing and feeding near the surface support vessels early in the project. Although observers on the surface had seen them diving at the site previously, they were not seen from *Sealab II* until day 26. Almost every night thereafter they appeared at *Sealab II* around 2000 hours. They were seen with increasing frequency, until in the final 2 weeks of the project at least 25 dives each night were observed. They commonly swam into the entryway of *Sealab II*, and on two occasions one surfaced in the open hatch. The latter events are of particular interest because the sea lion inhaled high-pressure gas and thus might well have experienced gas embolism upon its return to the surface. It must have exhaled during ascent, since it showed no signs of distress after returning to the surface.

The sea lions were seen eating anchovy, jack mackerel, white croaker, and squid. When a sea lion appeared, the fish would disperse from the ports; after it left, most fish gradually returned to the ports.

Discussion

The effect of the presence of *Sealab II* on the biota of the site area was almost immediate. Except for the octopus, the scorpion fish, and a few large predators, species of the sand-bottom community were notably absent from the site. One species, the pink sea perch, was obviously repelled by flashlights and wary of divers, but others—for example, the sand dabs—are not so wary and are known to be attracted to objects on the bottom. Although the smaller sand-bottom fishes were not found in the stomachs of predators, it is possible that some were present and were eaten by the larger predators aggregated around the site. It is also possible that the large numbers of scorpion fish on the bottom near *Sealab II* excluded them physically.

The most striking effect was the attraction of rocky-bottom and mid-water

Table 4. Standing crop of fishes for several natural and artificial habitats.

Type of reef or bottom	Area (m²)	Time on the bottom (days)	Number of resident fishes	Biomass (kg/m²)
Sealab II, California	500	43	5,800	2.86
Auto bodies, California*	360	910	10,900	?
Concrete blocks, Virgin Islands†	125	850	2,754	0.70
Coral, Virgin Islands†	600	Natural	1,352	.16
Coral, Bermuda‡	10,000	Natural		.049
Rock-kelp bed, Baja California§	1,850	Natural		.014
Sand near *Sealab II*		Natural		.082‖

* See *10*.   † See *11*.   ‡ See *13*.   § See *17*.   ‖ Includes estimates for scorpion fish and calico rockfish.

organisms to *Sealab II*. The tremendous increase in biomass of resident fishes (almost 35 times that of the normal sand-bottom community) is similar to results reported for artificial fishing reefs (Table 4). All the resident fishes at *Sealab II* were adults or near-adults. Larger fishes were also the first to appear at several artificial reefs off Southern California (*10*). Both of these observations contrast with observations from an artificial reef in the Virgin Islands (*11*), which was populated primarily by juvenile fishes.

Organisms were attracted to *Sealab II* far more rapidly than they have been to any artificial reef for which observations have been reported. This can in part be attributed to the proximity of a large reservoir of organisms in the canyon, but the light and noise around the site were doubtless important factors also. Since divers could sometimes detect the lights 20 meters away, it is possible that fish could detect them from as far away as the canyon; the noise from the site was probably detectable at greater distances. Thus, organisms within a fairly large area were probably aware of the presence of *Sealab II* almost immediately.

The success of artificial reefs in attracting marine organisms, especially fishes, has been attributed to the food and shelter available on the reef and also to simple thigmotaxis (*10*). There was no food growing on *Sealab II*, and, although there was abundant shelter for the small octopus, there was little for larger fishes. Fishes like the sheephead (*Pimelometopon pulchrum*) and the pile perch (*Rhacochilus vacca*) which live in the canyon but feed predominantly on benthic organisms were rare or absent. Species which apparently require shelter, such as cabezon and blacksmith, never became abundant. Only the scorpion fish were strongly thigmotactic; the remaining residents usually did not seek shelter or contact with solid objects. For them *Sealab II* was apparently a point of orientation on an otherwise featureless bottom. Because the residents remained around the site both day and night, their numbers increased during the project.

The visiting species appeared, for the most part, only at night and did not seek shelter or contact with solid objects. They were apparently attracted by the lights or prey around the lights and dispersed each morning as natural illumination increased. Consequently, their numbers did not increase during the project as did those of the substrate-oriented residents.

The visitors provided a regular source of food for the resident populations.

22

Because the lights attracted the zooplankton and, directly or indirectly, forage fishes from the surrounding water and concentrated them near the site each night, the amount of prey available at *Sealab II* was probably greater than that over an unlighted reef. Consequently, those species that were attracted to the substrate and able to utilize mid-water prey built up large resident populations. The situation was similar to a natural one reported off Baja California (*12*), where a steeply rising ridge intercepts an arm of the California Current. Although no observations were made by divers, investigations from the surface indicated a very large fish population on top of the ridge, apparently supported by vertically migrating zooplankton trapped on the ridge. In both cases the near-bottom fauna depends on a continually renewed and concentrated supply of mid-water organisms.

All of the resident fishes at *Sealab II* were carnivores, and 92 percent were secondary carnivores. The relative abundance of primary and secondary carnivores was apparently determined by the amount and type of attracted prey, since about 92 percent of the prey were small forage fish. Randall (*11*) and Bardach (*13*) also found a high percentage of carnivores in reef fish populations. In both cases many of the resident carnivores foraged over adjacent bottoms and not on the reef. These results illustrate the importance of nonresident food resources in determining the trophic structure of reef communities.

The number of mid-water organisms attracted at night appeared relatively constant, presumably because the lights drew from an approximately constant volume of water. If the supply of attracted organisms remained constant, the standing crop of residents would eventually be limited by a shortage of food. This apparently did not happen during the 44-day period, because the total number of residents and three of the four most abundant species were still increasing at the end of the project. The fourth species, vermilion rockfish, was obviously limited by something other than availability of foods; they reached a peak density early in the project and maintained it even though there apparently was interchange with canyon populations. By the end of the project the estimated biomass of prey attracted each night was about 2.8 percent of the standing crop of residents. This figure is close to daily food requirements that have been reported for fishes at similar temperatures (*14*). Even though our estimates of visiting prey are probably minimal, this figure suggests that the observed rate of increase in standing crop would not have continued much longer.

Seasonal and other long-term changes in the fauna would be expected. The species initially attracted to *Sealab II* were already present nearby in the canyon or along its edges. Undoubtedly other species would eventually wander in from more remote areas or appear as juveniles metamorphosing from planktonic larvae. Such new species could interact with those already present as prey, predators, or competitors. The disappearance of croakers after the sea lions arrived shows that the presence of even a single new species can drastically alter the density and trophic structure of the attracted fauna. It would be interesting to determine whether such marked changes would continue to occur as new species appeared, or whether, as predicted by theories of community diversity (*15*), the changes would become less violent as the diversity of the resident fauna increased. It would also be interesting to determine whether lighted reefs, because they appear to be more immediately effective in concentrating fishes, would also yield a greater sustainable harvest of commercial or sport fishes than unlighted reefs.

References and Notes

1. For details on Project Sealab II see *Man's Extension into the Sea* (Marine Technology Society, Washington, D.C., 1966); D. C. Pauli and G. P. Clapper, Eds., "Project Sealab Report, An Experimental 45-Day Undersea Saturation Dive at 205 Feet," *Office Naval Res. Rep. ACR-124* (1967).
2. E. W. Fager *et al., Limnol. Oceanog.* **11**, 503 (1966).
3. These fishes were *Hydrolagus colliei, Sardinops caerulea, Sphyraena argentea, Cynoscion nobilis, Pimelometopon pulchrum, Sebastodes serranoides, S. semicinctus, S. rubrivinctus, S. vexillaris,* and *Ophiodon elongatus.*
4. R. F. Ford, thesis, University of California, San Diego, 1965.
5. We use the term *school* to describe only those aggregations in which most of the individuals are oriented in the same direction.
6. Attracted fishes that appeared later and in small numbers were *Palometa simillima, Seriphus politus, Embiotoca jacksoni,* and *Sebastodes jordani.*
7. C. Limbaugh, *Univ. Calif. Inst. Marine Resources Pub. IMR ref 55-9* (1955), pp. 50-156.
8. T. Eisner, R. Alsop, G. Ettershank, *Science* **146**, 1058 (1964).
9. P. B. Taylor, thesis, University of California, San Diego, 1963.
10. J. G. Carlisle, C. H. Turner, E. E. Ebert, *Calif. Fish and Game, Fish Bull.* **124** (1964), p. 8.
11. J. E. Randall, *Caribbean J. Sci.* **3**, 31 (1963).
12. J. D. Isaacs and R. A. Schwartzlose, *Science* **150**, 1810 (1965).
13. J. E. Bardach, *Limnol. Oceanog.* **4**, 77 (1959).
14. G. Thorson, *Perspectives in Marine Biology,* A. A. Buzzati-Traverso, Ed. (Univ. of California Press, Berkeley, 1958), pp. 70-71.
15. C. S. Elton, *The Pattern of Animal Communities* (Methuen, London, 1966), pp. 377-78; R. S. MacArthur, *Ecology* **36** (1955).
16. Rare sand-bottom species not listed in Table 2 were sea stars, *Henricea leviuscula* and *Pycnopodia helianthoides;* crabs, *Cancer productus* and *Loxorhynchus* sp.; snail, *Megasurcula carpenteriana;* fishes, *Torpedo californica, Hippoglossina stomata, Microstomus pacificus, Otophidium taylori,* unidentified agonids, and *Porichthys notatus.*
17. J. C. Quast, *Univ. Calif. Inst. Marine Resources Pub. IMR ref 60-3* (1960), p. 22.
18. The study discussed here was supported by the Office of Naval Research NONR 2216(30). We thank all personnel of Project Sealab II for their cooperation, especially Commander M. S. Carpenter and the other aquanauts for their invaluable aid during our stays on the bottom. We also thank Miss Thea Schultze for technical assistance, Dr. E. W. Fager for his aid and counsel, and Dr. R. H. Rosenblatt for critical reading of the manuscript. This article is a contribution from the Scripps Institution of Oceanography, University of California, San Diego.

THE EFFECT OF ENVIRONMENTAL TEMPERATURE
ON THE FATTY ACID COMPOSITION OF
CRUSTACEAN PLANKTON

Tibor Farkas and Sandor Herodek

FISH FATS ARE characterized by high proportions of C_{20-22} highly unsaturated fatty acids. It was demonstrated as early as 1932 (1) that marine fish contain significantly more C_{20-22} than do fresh-water fish. From that time on the "fresh-water" and "marine" indications came into general use for types of fatty acid composition, but no explanation has been offered for this difference. The question is in close correlation with the origin of C_{20-22} fatty acids present in fish.

Recently isotope studies (2) demonstrated the synthesis in fish of C_{20-22} polyenoic acids from exogenous precursors of linoleic and linolenic types; however, earlier feeding experiments (3) indicated that the greater majority of such polyenoic acids derived from food. If this is the case, the problem is why the food of marine fish contains greater amounts of C_{20-22} than that of fresh-water fish. The food chain of water ecosystems can be simplified to algae, crustaceans, and fish. Some data (4) have been obtained from one marine species and a small sample of fresh-water species indicating that differences exist in the fatty acid composition of marine and fresh-water crustaceans similar to those of marine and fresh-water fish. The problem of the two types of fatty acid composition has now been shifted from the level of fish down to crustaceans. The authors investigated previously the fatty acid composition of the fresh-water crustacean plankton (5, 6) and found that with lowering of the environmental temperature the amount of C_{20-22} gradually increases and the fatty acid composition becomes similar to that of marine animals.

This observation led to the assumption that the environmental temperature is one of the factors responsible for the differences in fatty acid composition between marine and fresh-water animals. In this connection data are presented here for crustaceans sampled from cold and warm seas. The authors' previous observations were made on the mixed crustacean plankton of a fresh-water lake. Now the seasonal cycle of the fatty acid composition of several selected species will be demonstrated. The biological significance of the high amount of C_{20-22} acids in crustaceans needs some interpretation. It seemed probable that these highly unsaturated acids may have a role in assuring the liquid state of the stored lipids. This hypothesis is investigated now by comparing the melting point of lipids with the environmental temperature in different seasons. As to the origin of C_{20-22} acids present in crustaceans some uncertainty remained after the previous work. In the few investigated unicellular alga species these acids are never present in significant amounts, but it is practically impossible to control the fatty acids of all the organisms which could serve as food of crustaceans under natural conditions. It seemed therefore desirable to investigate this problem in feeding experiments under laboratory conditions.

MATERIALS AND METHODS

The planktonic crustaceans are small animals, a few millimeters in length. They obtain their food, consisting of unicellular algae, chiefly by filtration. The main body of crustacean plankton is composed of two groups of lower crustaceans. These are the cladocerans and copepods, two groups which show significant differences in their morphology, behavior, and distribution. The copepods are represented both in fresh water and sea by a large number of species and individuals, while the role of cladocerans is inferior in the sea.

The mixed crustacean plankton was collected with a No. 6 net. First we analyzed it as a whole. Later we succeeded in separating the two main groups, the cladocerans and copepods. After the water containing the collected material had been vigorously aerated, the cladocerans—all of them possessing a hydrophobic cuticula—adhered to the surface and could be quantitatively removed. The isolation of the single species in each case required laborious procedures, worth trying only if the species desired was abundant in the sample. In this manner, we obtained from the mixed crustacean plankton 30–50 mg, and from the selected species 1–3 mg, of fat. The marine crustaceans we received in glass vials, filled with 70% ethanol, sealed after flushing with CO_2. The crustacean fed on algae were kept in 100-liter aquaria. The algae originated from pure cultures. Under these conditions the cladocerans propagated quickly and many generations were fed on algal diet before analysis. The copepods were kept on algae from the nauplius stage on and were analyzed when adult. Our experiments with fish started with newly born guppies, *Lebistes reticulatus*. They were raised for 2 months on living crustacean plankton, from Lake Balaton, as their only food. Every 3rd day, fresh crustaceans were added in a quantity just sufficient to last until the next feeding.

All samples were ground with anhydrous sodium sulfate and extracted three times with petroleum ether (boiling range 40–70°), for 15 min under reflux, using 30 ml petroleum ether per g sample. The extraction and subsequent procedures were carried out in a strictly inert atmosphere. The lipids were dissolved in petroleum ether and stored in a refrigerator at $-20°$, sealed in glass vials after flushing with CO_2.

On an aliquot of the total lipid fraction, we measured the iodine value, using Kaufmann's semimicro method (7). Another portion was hydrogenated in the presence of a palladium catalyst. The fatty acid mixture obtained after hydrogenation was analyzed by paper chromatography according to Kaufmann (7). To detect the spots, copper acetate–rubeanic acid reagent was used. The quantitative evaluation was performed photometrically. The results were converted into percentages

27

by weight. We determined the accuracy of this technique several times by the use of model fatty acid mixtures and the results corresponded to those postulated by Seher (8). Each sample was chromatographed on three independent occasions and the deviations never exceeded 5%.

For melting point measurements we employed the Kofler-type microscope, with a temperature-conditioned sample holder. A 2-3 mg sample of fat was placed on the slide, cooled to a temperature 20° below the expected melting point and afterward heated at a speed of 1° per min. The melting point was taken as the temperature at which the minute roughness of the surface of the fat droplet disappeared. This change was rapid and easily observed and the results from this technique were reproducible to within ±1°.

RESULTS

As a preliminary assay of the suspected seasonal changes, iodine value measurements were performed on the fat of crustacean plankton. The results are presented in Fig. 1. The equation of the regression line calculated from these data is:

$$\text{Iodine value} \quad (\text{I.V.}) = 188.8 - (3.03 \pm 0.21)t.$$

The deviation of the regression coefficient from 0 is very highly significant ($P < 0.001$).

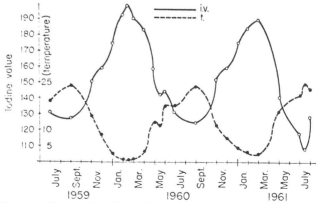

Fig. 1.   Changes in iodine value of the fat of the crustacean plankton and of the water temperature in Lake Balaton, 1959-61.

It seemed probable that the chemical changes indicated by the iodine values result in a relatively constant physical state of the planktonic crustacean fat throughout the year. In order to characterize this state, melting point measurements were carried out on the fat of planktonic copepods sampled from Lake Balaton in 1962. The results were as follows (date, mean water temperature in period immediately preceding measurement, mp): 9 August, 22.8°, 19°; 11 September, 20.3°, 17°; 9 October, 16.2°, 12°; 15 October, 15.3°, 12°; 29 October, 15.1°, 11°; 6 November, 9.4°, 8°; 13 November, 9.2°, 5°; 4 December, 2.6°, 2°. It should be pointed out that in these crustaceans the fat is stored extracellularly in the so-called oil sacs. We believe that the physical state of the fat, stored in this way, corresponds fairly well to that observed under the microscope, through the structural lipids, also extracted, may cause some smaller deviations. It appears that the melting points of the fats of planktonic crustaceans are always just below the environmental temperature.

In some previous work (5, 6), paper chromatography of the hydrogenated fatty acids of crustacean plankton showed in one cooling period of Lake Balaton an increase of the proportion of C_{20-22} acids. Similar observations made in two subsequent years are shown together with the earlier results in Table 1.

The gas–liquid chromatography (GLC) analysis of one of the samples revealed the following composition (as percentages of the total peak areas).[1] $14:0 = 22.5\%$, $16:0 = 11.6\%$, $16:1 = 2.5\%$, $16:2 = 0.4\%$, $18:0 = 0.7\%$, $18:1 = 2.9\%$, $18:2 = 9.8\%$, $18:3 = 9.0\%$, $18:4 = 14.4\%$, $18:? = 2.7\%$, $20:4 = 4.3\%$, $20:5 = 9.6\%$, $22:5 = 2.3\%$, $22:6 = 7.2$. According to these more detailed data, the C_{20-22} acids of crustacean plankton are indeed tetra-, penta- and hexaenoic acids, a result analogous to that obtained from an analysis of the C_{20-22} acids of fish (9).

---

[1] The apparatus employed was a Wheelco Model 10 (Barber-Colman Co.) with a 6 ft, 6 mm i.d. column of ethylene glycol succinate polyester, 16% on 80–120 mesh silanized Chromosorb at 185–200°. Compounds were identified by comparison of their retention times with those of known standards. Areas were measured by triangulation. Fatty acids are designated by chain length and number of double bonds.

TABLE 1   SEASONAL CHANGES IN THE FATTY ACID COM-
POSITION OF THE MIXED CRUSTACEAN PLANKTON OF LAKE
BALATON

Date of Sampling	Water Temperature	Fatty Acid Composition*				
		C_{14}	C_{16}	C_{18}	C_{20}	C_{22}
	°			%		
1959 July 14	23.0	4.5	25.5	38.7	16.0	15.3
Aug. 24	23.5	1.0	27.6	39.7	15.5	15.8
Oct. 4	12.5	5.4	20.5	37.7	20.6	15.8
Nov. 14	7.2	8.1	18.2	32.6	20.6	20.5
Dec. 14	2.3	8.1	14.0	38.9	17.6	21.4
1960 June 1	17.0	5.7	26.2	37.1	17.7	13.3
Aug. 29	24.5	5.4	25.5	34.4	13.3	11.4
Oct. 24	13.5	7.8	22.0	37.0	16.2	17.0
Dec. 20	4.5	7.7	12.0	35.0	21.7	23.6
1961 June 13	20.5	2.6	44.6	38.2	9.5	5.1
July 3	24.9	—	37.2	43.5	11.8	7.5
July 13	22.6	1.4	29.0	42.1	12.8	14.7
Oct. 11	15.4	7.2	28.2	35.5	16.6	12.5
Nov. 10	9.0	13.1	8.5	34.7	21.2	22.5
Nov. 21	7.6	7.1	15.1	38.3	18.1	21.4

* After hydrogenation.

Recently we succeeded in selecting some individual crustacean species in quantities large enough for analysis. The results are presented in Table 2. The fatty acid composition of each species proved to be different and was modified differently by the changes of water temperature. No C_{20-22} acids were detectable in *Daphnia magna* and the amount of these acids in *Daphnia cucullata* showed only a slight increase as the temperature decreased. The *Daphniae* belong to the cladocerans. On the other hand in the two copepod species *Eudiaptomus gracilis* and *Cyclops vicinus* there appeared to be a definite increase in C_{20-22} acids at lower temperatures. From a biological point of view these two copepod species are better adapted to the cold environment than the two cladoceran species.

The (hydrogenated) fatter acid compositions of two marine copepod species from different seas proved to be:

30

TABLE 2 Seasonal Changes in the Fatty Acid Composition of Selected Crustacean Species

Species	Place	Date	Water Temperature	C_{14}	C_{16}	C_{18}	C_{20}	C_{22}
Eudiapt. gracilis (Copepoda)	Lake Balaton	June 62	22.5	—	34.2	35.5	16.8	13.5
		Sept. 62	18.0	7.4	29.9	28.5	15.4	18.8
		Nov. 62	10.1	16.4	20.8	30.9	12.9	19.0
		Nov. 62	2.8	9.7	13.8	36.9	19.9	19.7
Cyclops vicinus (Copepoda)	Lake Balaton	Oct. 62	16.2	11.5	29.9	27.3	13.7	17.6
		Oct. 62	15.4	7.8	27.6	28.1	17.7	18.8
		Nov. 62	9.4	10.7	22.8	25.9	16.7	23.9
		Nov. 62	7.0	8.4	16.7	33.3	16.2	25.4
		Nov. 62	2.8	6.1	15.6	36.3	16.2	25.8
Cyclops vicinus (Copepoda)	Lake Belsötö	Sept. 60	23.5	—	20.5	37.0	19.4	23.1
		Oct. 60	13.9	—	22.2	24.9	21.2	31.7
		Dec. 60	5.0	—	18.4	22.7	23.0	35.9
		June 61	20.5	—	37.0	29.3	15.0	18.7
		July 61	21.5	—	21.9	40.9	21.6	15.6
		Nov. 61	9.0	—	29.3	30.3	19.6	20.8
Daphnia cucullata (Cladocera)	Lake Balaton	June 61	24.0	9.0	36.4	42.6	15.8	5.2
		Aug. 61	16.2	11.5	31.6	39.7	14.8	4.9
		Oct. 61	15.4	13.4	28.9	38.9	15.3	5.4
		Oct. 61	9.4	14.7	25.4	40.5	11.7	9.0
		Nov. 61	9.2	16.3	22.8	42.9	9.6	10.0
		Nov. 61	0.0	10.8	17.3	42.2	15.0	9.2
		Feb. 62			22.9	40.7	13.3	12.3
Daphnia magna (Cladocera)	Lake Belsötö	June 61	20.5	—	57.3	35.0	7.7	—
		Nov. 61	9.0	—	37.5	54.7	7.8	—

* After hydrogenation.

31

Paracalanus parvus from Bay of Naples: $C_{14} = 6.8\%$, $C_{16} = 56.3\%$, $C_{18} = 27.2\%$, $C_{20} = 4.7\%$, $C_{22} = 5.0\%$. *Calanus finmarchicus* from the North Sea: $C_{14} = 6.8\%$, $C_{16} = 29.4\%$, $C_{18} = 22.0\%$, $C_{20} = 23.6\%$, $C_{22} = 18.2\%$.

It is too early to consider the problem settled on the basis of this single comparison, but the results indicate that the temperature effect is as easily detectable in different seas as it is during a seasonal temperature change in a fresh-water lake.

As a first step in clarifying the mechanism of the temperature effect, the origin of the fatty acids present in crustaceans must be investigated. For this reason we started with feeding experiments under controlled conditions (See Table 3). Three Cladocera and three Copepoda species were raised in aquaria on unicellular green algae *Scenedesmus obtusiusculus*. The fat of the cladocerans remained similar to that of their food. In the copepods, on the other hand, the C_{20-22} acid group appeared always as the major component. It seems plausible that the copepods obtained these acids by elongation of the linoleic and linolenic acids according to the divinyl methane rhythm, as demonstrated in vertebrates (10, 11). The temperature effect was also investigated under aquarium conditions, but for technical reasons only on *Daphnia magna* (see again Table 3). Two groups were kept at 20° and 10°, respectively. In this case *Chlorella pyrenoidosa* (green algae) served as food. Again there were no C_{20-22} acids in the *Daphnia magna*, but at the lower temperature the proportion of C_{18} increased. The iodine value difference between the crustaceans at the two temperatures corresponded to the equation given above for natural conditions, though the absolute values are somewhat higher. We assume that all planktonic crustaceans react to decreasing temperature by accumulation of polyenoic acids, but in some species the C_{18}, rather than the C_{20-22}, polyenoic acids are of greater importance in this process.

The characteristic fatty acid compositions of different planktonic crustaceans permit a simple and natural test of the effect of diet on the fats of fish. In aquaria at 20° two groups of guppies were raised on summer and winter plankton respectively. The fatty acid compositions were as follows:

TABLE 3 FEEDING EXPERIMENTS WITH WATER ORGANISMS

Sample	Aquarium Temperature °	Iodine Value	Fatty Acid Composition* %				
			C₁₄	C₁₆	C₁₈	C₂₀	C₂₂
Algal diet (*Scenedesmus obtusiusculus*)	—	—	—	47.4	52.6	—	—
Cladocerans:							
Daphnia magna	ca. 20	—	—	42.5	57.4	—	—
Daphnia cucullata	20	—	—	44.4	52.4	3.2	—
Moina rectirostris	20	—	—	46.3	50.8	2.9	—
Copepods:							
Mesocyclops leuckarti	20	—	—	39.9	42.4	7.7	10.0
Cyclops vicinus	20	—	6.9	19.4	39.8	18.3	15.6
Acanthocyclops viridis	20	—	—	32.0	30.0	20.2	17.8
Algal diet (*Chlorella pyrenoidosa*)	20	121.5	—	35.7	64.3	—	—
Daphnia magna	25	127.7	—	38.9	61.1	—	—
Daphnia magna	10	173.2	—	32.6	67.4	—	—
Fish fed on crustacean plankton:							
Lebistes reticulatus							
Summer	22		6.4	31.9	27.8	12.2	21.7
Winter	22		4.2	14.7	20.0	25.3	35.8

* After hydrogenation.

33

Fish fed on summer plankton: $C_{14} = 6.4\%$, $C_{16} = 31.9\%$, $C_{18} = 27.8\%$, $C_{20} = 12.2\%$, $C_{22} = 21.7\%$. Fish fed on winter plankton: $C_{14} = 4.2\%$, $C_{16} = 14.7\%$, $C_{18} = 20.0\%$, $C_{20} = 25.3\%$, $C_{22} = 35.8\%$.

In this experiment, the seasonal difference detected in the crustacean plankton manifests itself clearly in fish, although the quantity of C_{20-22} was higher in the fat of both groups of fish than in the dietary fat.

DISCUSSION

On the basis of the above results, it is suggested that the main pathway for the biogenesis of the C_{20-22} fatty acids in aqueous ecosystems is as follows. The algae produce large amounts of C_{16-18} polyenes. The crustaceans, mainly the copepods, elongate these acids to C_{20-22}. The fish readily take up the largest part of the C_{20-22} fatty acids from the crustaceans and store it. The significance of the chain elongation for the crustaceans is that the C_{20-22} highly unsaturated acids ensure a low melting point for the extracellularly stored lipids. With reference to the lipids in fish, we believe that it is not the salinity of the body of water, but its size that is the determining factor. The characteristic fatty acid composition accepted for marine fish is based on species caught in the comparatively cold North Sea, where the fat of the crustacean plankton must be very rich in highly unsaturated fatty acids. As one consequence of smaller dimensions, most fresh-water lakes become considerably warmer in summer. As the temperature decreases later in the year, the feeding activity of fish declines and even stops, so that their food, consisting largely of summer plankton, is very poor in the longer-chain fatty acids. Our recent results on marine crustaceans and the data found in the literature concerning fish from warm seas (12) fit into the picture we have already suggested (13). As a second consequence of the smaller dimensions in fresh water the importance of the food originating on land, shore, and bottom increases, and this kind of food is poor in polyenes.

The different distribution of the cladocerans and copepods in fresh water and the seas might be a further factor contributing to the formation of marine and fresh-water fatty acid types of fish oils (13).

We are greatly indebted to Professor J. F. Mead, University of California Medical Center, Los Angeles, for his kind interest and manifold help in our work; to Professor S. M. Marshall, Marine Station, Millport, Isle of Cumbrae, Scotland, to Dr. A. Packard, Zoological Station, Naples, Italy for sending us the marine crustaceans; and to Mrs. Vida Slauson of The Lipid Research Laboratory, Department of Biological Chemistry, University of California at Los Angeles, for performing gas–liquid chromatographic analyses on some of the samples.

References

1. Lovern, J. A. *Biochem. J.* **26:** 1978, 1932.
2. Mead, J. F., M. Kayama, and R. Reiser. *J. Am. Oil Chemists' Soc.* **37:** 438, 1960.
3. Kelly, P. B., R. Reiser, and D. W. Hood. *J. Am. Oil Chemists' Soc.* **35:** 503, 1958.
4. Hilditch, T. P. *The Chemical Constitution of Natural Fats.* Chapman and Hall Ltd., London, 3rd ed. 1956, p. 30.
5. Farkas, T., and S. Herodek. *Acta Biol. Acad. Sci. Hung.* **10:** 85, 1959.
6. Farkas, T., and S. Herodek. *Magy. Tud. Akad. Tihanyi Biol. Kutatoint. Evkonyve* **28:** 91, 1961.
7. Kaufmann, H. P. *Fette, Seifen, Anstrichmittel* **56:** 154, 1954.
8. Seher, A. *Fette, Seifen, Anstrichmittel* **61:** 855, 1959.
9. Insull, W., and E. H. Ahrens, Jr. *J. Lipid Res.* **1:** 199, 1960.
10. Mead, J. F. *Am. J. Clin. Nutr.* **8:** 55, 1960.
11. Klenk, E., and H. Mohrhauer. *Z. Physiol. Chem.* **320:** 218, 1960.
12. Karkhanis, Y. D., and N. G. Magar. *J. Am. Oil Chemists' Soc.* **32:** 492, 1955.
13. Farkas, T., and S. Herodek. *Magy. Tud. Akad. Tihanyi Biol. Kutatoint. Evkönyre* **29:** 79, 1962.

ENOLASE ACTIVITY IN MARINE PLANKTONIC ALGAE

N. J. Antia, J. Kalmakoff, and A. Watt

Introduction

The occurrence of the Embden–Meyerhof glycolytic pathway of carbohydrate metabolism in most forms of life examined has led to a general ex-
pectation that enzymes of this pathway operate in many types of organisms
not yet studied. This has resulted in few attempts to demonstrate the occurrence
of even the key enzymes of this pathway in a very heterogeneous group of
organisms, the algae (1). In particular, the enzyme enolase (2-phospho-D-gly-
cerate hydro-lyase (EC 4.2.1.11)) does not appear to have been examined in
any form of alga.

The availability of a number of marine planktonic algal species in pure
culture has offered the opportunity to make a preliminary examination of
enolase activity in members of the Bacillariophyceae, Chlorophyceae, Chrysophyceae, Cryptophyceae, Dinophyceae, and Myxophyceae. The results of such
an examination of algal cells collected from axenic mass culture of 14 species
are presented in this paper. Owing to the extensive size of this survey and the
limited quantity of cellular material obtained, the examination was restricted
to crude cell-free extracts prepared by one generally applicable method that
appeared to be most advantageous. The method of cell disruption by treatment
with cold acetone followed by buffer extraction of the resultant acetone
powders was chosen for the reason previously given (2).

Materials and Methods

Algal Species

The species studied are listed in Table I as members of the algal class to
which they are known to belong. All the species were obtained as specimen
cultures from algal collections of other laboratories (3).

Mass Culture Methods

The apparatus, culture media, and methods used for culture initiation,
growth measurement, check on axenic condition, cell harvest, and storage have
been previously described (3). The checks made indicated that all the cultures
were free from microbial contamination up to the time of cell harvest. To
obtain maximum yields of metabolically active "young" cells, all cultures were
harvested towards the end of (or soon after) the logarithmic phase of growth.

Acetone Powders

These were prepared by a standard procedure similar to that previously described (2).

Cell-Free Algal Extracts

Extracts of *Dunaliella tertiolecta, Monochrysis lutheri, Isochrysis galbana*, and species of Bacillariophyceae were prepared from the acetone powders with sodium orthophosphate buffer (0.019 M, pH 7.5) exactly as described before (2). With the other algal species, the strength of the phosphate buffer was increased to 0.05 M and the extraction at 0–4 °C was prolonged to 16 hours.

Enolase Determinations

All tests for enolase activity were based on the spectrophotometric method introduced by Warburg and Christian (4). The substrate, tricyclohexylammonium-D-glyceric acid 2-phosphate, was kindly provided by Dr. H. Tsuyuki (5). Crystalline muscle enolase and tricyclohexylammonium phosphoenolpyruvate were purchased from Sigma Chemical Company, and Cleland's reagent (dithiothreitol) from Calbiochem.

Test Medium

A medium based on that of Holt and Wold (6) was chosen for the initial standard tests on all algal preparations. This medium (A) was composed of 1 mM substrate, 1 mM MgSO$_4$, 200 mM KCl, and 50 mM imidazole, with the pH adjusted to 7.0 with HCl. For the tests intended to ensure absence of possible heavy-metal inhibition, this medium was modified in the manner of Wood (7) to include a low concentration of ethylenediaminetetraacetate (EDTA) (0.01 mM) and the KCl concentration was increased to 400 mM. When —SH group protection was desired, 10–20 μmoles of Cleland's reagent (8) was incorporated into 3 ml of either medium. All media were periodically checked for suitability in enzyme tests by control experiments with standard solutions of muscle enolase.

Test Methods

Enzyme activity was tested in algal extracts and acetone powders. Every extract showing activity was further tested for enzyme inhibition by fluoride–phosphate. All tests were carried out at room temperature (20–25 °C).

To test the algal extract, 3 ml of medium was put into a cuvette (1-cm light path), 0.1 ml of the extract was added with a plexiglass ladle, rapidly mixed in, and the optical density at 240 mμ (O.D.$_{240}$) was read immediately in a Beckman DU spectrophotometer. O.D.$_{240}$ readings were taken continually at suitable intervals thereafter for at least 4 hours or until no significant change was observed. No attempt was made to arrive at equilibrium in all cases. Simultaneous control measurements were made on the extract in the same medium *without* substrate. After incubation for 4–5 hours (or after equilibrium), the ultraviolet absorption of each test mixture was scanned against its control in a Beckman DB recorder-fitted spectrophotometer; a typical absorption curve obtained is shown in Fig. 2(C). Fluoride–phosphate inhibition was similarly tested after the incorporation of 0.01 ml of a solution containing 0.3 M NaF and 1 M KH$_2$PO$_4$ into 3 ml of the medium before addition of the extract.

To test acetone powder, a suspension of 7 mg of the powder in 4 ml of medium was shaken gently on a Virtis extractomatic machine for 2 hours, then centrifuged for 15 minutes at about 12,000 g. The clear supernate was put into a cuvette, its O.D.$_{240}$ was measured, and its ultraviolet absorption was scanned as described above. A control was used with each acetone powder taken in the medium *without* substrate.

Calibration

Aliquots (0.1 ml) of standard solutions of phosphoenolpyruvate (PEP) were incorporated into 3-ml aliquots of medium A, and O.D.$_{240}$ values were measured to give the calibration curve shown in Fig. 1(A). The conversion factor obtained from this curve (viz. 0.439 unit increase in O.D.$_{240}$ value per μmole PEP added) was used to calculate enzyme activity in algal extracts. The amount of enzyme catalyzing the formation of 1 μmole PEP per hour under the test conditions was chosen as the unit of activity. The activity values reported in Table I were calculated from those periods of incubation (generally the first 10 minutes) during which maximum steady rate of increase in O.D.$_{240}$ values was observed (Figs. 3 and 4).

Protein Determinations

Algal extracts were diluted 25- to 50-fold, and protein was determined in the diluted extracts by a procedure based on the method of Lowry (9). Suitable blanks were used to correct for the color of extracts. Crystalline bovine-serum albumin (Sigma) and rabbit-muscle aldolase (Sigma) were used as standards.

Results and Discussion

The results of standard tests on the algal extracts that gave significant change in O.D.$_{240}$ during the first hour of incubation are depicted in Figs. 3 and 4. An example of similar control tests made with muscle enolase is shown in Fig. 1(B). The reaction kinetics observed are indicative of enolase activity in these species of algae. The tests on acetone powders also indicated activity of a degree comparable to that observed in the corresponding extracts. The ultraviolet absorption spectra of the test products gave further proof of the formation of PEP; the absorption maximum at about 232 mμ characteristic of PEP was observed with products from all algal extracts showing activity. Typical examples of the ultraviolet absorption of PEP and of reaction products from tests made with muscle enolase and an algal extract are shown in Fig. 2.

A marked inhibition of reaction rate by fluoride–phosphate was observed with all active algal extracts (Figs. 3 and 4). The inhibition was verified with muscle enolase for the test conditions used (Fig. 1(C)), and was confirmed by the ultraviolet spectrum of the inhibition-test products (e.g. Fig. 2(C)). Since fluoride–phosphate inhibition is characteristic of enolase from other organisms hitherto studied (10–12), the observed inhibition offers further

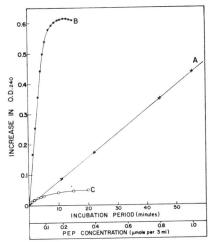

FIG. 1. Enolase test calibration and controls (PEP = phosphoenolpyruvate). A: increase in O.D.$_{240}$ with PEP concentration; B: typical control reaction from 0.1 ml muscle enolase standard (40 μg/ml, protein of specific activity 3420 units) in 3 ml test medium A (6); C: as for B, with fluoride–phosphate added.

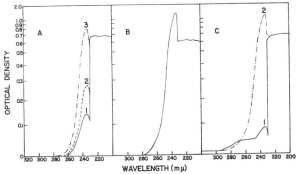

FIG. 2. Ultraviolet absorption spectra of enolase test products scanned against test medium. A: curve 1, 0.2 μmole, curve 2, 0.4 μmole, and curve 3, 1.0 μmole of PEP in test medium; B: equilibrium product from muscle enolase control corresponding to Fig. 1(B); C: equilibrium product from reaction of 0.1 ml *Cryptomonas* sp. extract, curve 1, with fluoride–phosphate, and curve 2, without fluoride–phosphate.

evidence for enolase activity in the algal extracts and suggests a similar requirement of metal ions for activity.

The values in Table I reveal that high specific activities (10–20 units) were obtained from members of the Chlorophyceae and a diatom species in relation to the activity level generally observed (2–4 units). The particularly low activity obtained from *H. virescens* may be due to partial cell lysis unexpectedly encountered during washing of cells after harvest. There is doubt about the quantitative significance of the observed activity differences, since no attempt was made to ensure complete enzyme extraction and the standard methods

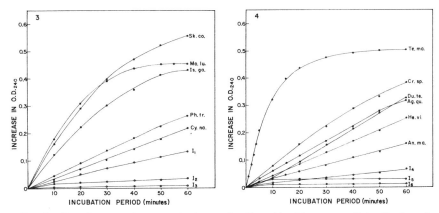

Fig. 3. Reaction rates from standard tests (medium A) with extracts (0.1 ml) of species of Bacillariophyceae and Chrysophyceae. Specific rate curves are identified by species name initials (see Table I). Fluoride–phosphate inhibitions are shown by I_1 for *M. lutheri*, I_2 for *S. costatum*, and I_3 for *C. nana*, *I. galbana*, and *P. tricornutum*.

Fig. 4. Reaction rates from standard tests (medium A) with extracts (0.1 ml) of species of Chlorophyceae, Cryptophyceae, and Myxophyceae. Specific rate curves are identified by species name initials (see Table I). Fluoride–phosphate inhibitions are shown by I_4 for *T. maculata*, I_5 for *Cryptomonas* sp. and *A. marina*, and I_6 for *D. tertiolecta*, *H. virescens*, and *A. quadruplicatum*.

TABLE I

Protein content and enolase activity of algal extracts

Alga	Protein (mg/ml)	Enolase activity	
		Units/ml	Units/mg protein
Bacillariophyceae			
Phaeodactylum tricornutum	2.3	6.3	2.7
Skeletonema costatum	2.0	21.9	10.9
Cyclotella nana	1.4	4.7	3.4
Chlorophyceae			
Dunaliella tertiolecta	0.7	8.2	11.7
Tetraselmis maculata	3.0	57.5	19.2
Chrysophyceae			
Monochrysis lutheri	5.8	24.6	4.2
Isochrysis galbana	4.1	16.4	4.0
Coccolithus huxleyi	3.6	n.m.*	—
Cryptophyceae			
Cryptomonas sp. (3 C)	3.5	10.9	3.1
Hemiselmis virescens	7.7	5.8	0.8
Rhodomonas lens	2.6	n.m.	—
Dinophyceae			
Amphidinium carteri	2.2	n.m.	—
Myxophyceae			
Agmenellum quadruplicatum	3.6	8.2	2.3
Anacystis marina	1.5	3.6	2.4

*Not measurable.

used would be expected to show considerable species discrimination. Apart from indicating the occurrence of substantial enolase activity in these algal species, the data obtained are inadequate for quantitative consideration of the activity in vivo.

No activity was detected in extracts or acetone powders of *C. huxleyi, R. lens,* and *A. carteri* when they were tested by the standard methods. In each case, the subsequent incorporation of muscle enolase into the test system gave the usual activity associated with this enzyme, indicating that the failure to detect activity was not due to suppression from the presence of usual enolase inhibitors. The protein concentrations of the extracts were of the usual order of magnitude obtained with the other algal species (Table I). *C. huxleyi* could not be examined further on account of shortage of cell material. *R. lens* and *A. carteri* were reexamined under conditions known to prevent possible inhibition by heavy metals (with the use of EDTA), and to protect a possible SH-containing enzyme in which the —SH is sensitive, such as potato enolase (11). Dithiothreitol was chosen as protective reagent for —SH groups because of its obvious advantages (8) and was found to show no interference in control tests made with muscle enolase. With EDTA addition, *R. lens* showed very low activity, detectable on prolonged incubation but hardly measurable in the first hour. The activity was not enhanced by incorporation of dithiothreitol. It appears that the enzyme present may have unusual requirements for a satisfactory demonstration of activity. On the other hand, *A. carteri* gave no indication of activity under all conditions tested. The failure to detect aldolase activity as well in this species (unpublished observation) suggests a unique enzymatic "make-up" in this organism, probably involving an unorthodox pathway of carbohydrate metabolism.

Acknowledgments

The kindness of Dr. H. Tsuyuki in providing a generous quantity of the substrate and in offering some useful advice is gratefully appreciated.

References

1. G. JACOBI. *In* Physiology and biochemistry of algae. *Edited by* R. A. Lewin. Academic Press, Inc., New York and London. 1962. p. 125.
2. N. J. ANTIA and A. WATT. J. Fisheries Res. Board Can. **22**, 793 (1965).
3. N. J. ANTIA and J. KALMAKOFF. Fisheries Res. Board Can. Manuscript Rept. Ser. Oceanog. Limnol. No. 203 (1965).
4. O. WARBURG and W. CHRISTIAN. Biochem. Z. **310**, 384 (1941–1942).
5. H. TSUYUKI and F. WOLD. Science, **146**, 535 (1964).
6. A. HOLT and F. WOLD. J. Biol. Chem. **236**, 3227 (1961).
7. T. WOOD. Biochem. J. **91**, 453 (1964).
8. W. W. CLELAND. Biochemistry, **3**, 480 (1964).
9. R. D. DEMOSS and R. C. BARD. *In* Manual of microbiological methods. *Edited by* Committee on Bacteriological Technic of Society of American Bacteriologists. McGraw-Hill Book Co., Inc., New York. 1957. p. 181.
10. O. MEYERHOF. *In* The enzymes. Part 2. 1st ed. Vol. I. *Edited by* J. B. Sumner and K. Myrbäck. Academic Press, Inc., New York. 1951. p. 1215.
11. H. BOSER. Z. Physiol. Chem. **315**, 163 (1959).
12. R. A. MACLEOD, R. E. E. JONAS, and E. ROBERTS. Can. J. Biochem. Physiol. **41**, 1971 (1963).

BIOASSAY OF SEAWATER

I. A ¹⁴C UPTAKE METHOD FOR THE DETERMINATION OF CONCENTRATIONS OF VITAMIN B₁₂ IN SEAWATER

A. F. CARLUCCI AND S. B. SILBERNAGEL

Introduction

The B_{12} requirement of many phytoplankton (4, 7) has necessitated the development, for quantitative marine ecology, of a method for the rapid determination of this vitamin in large numbers of seawater samples. The very low concentrations of the vitamin in seawater require a method sensitivity which can only be approached by bioassays. Both bacteria and algae have been employed in the bioassay of seawater for vitamin B_{12} (1, 7). These methods involve 1- to 21-day incubation periods, generally the time required for terminal growth of the organisms used, followed by the determination of cell numbers or optical densities. The range of B_{12} concentrations measured by these methods is from less than 0.1 to about 20 $\mu\mu g$ B_{12} per milliliter. Ryther and Guillard (8) and Menzel and Spaeth (6) used *Cyclotella nana* Hustedt in seawater bioassays in which cell numbers were determined. More recently Gold (2) described a technique for B_{12} bioassay using *C. nana* Hustedt, in which the ¹⁴C incorporated during photosynthesis was used to measure the alga's response to B_{12} concentrations after a 24-hour incubation. The range of the method was 0.1 to 6.0 $\mu\mu g$ B_{12} per milliliter. A modified Gold method is described which employs ¹⁴CO₂ incorporation in *C. nana* after an approximately 48-hour incubation, a better period for allowing the cells to adapt to a variety of seawaters. In samples from the Pacific Ocean serious inhibition of the alga was noted, a factor not taken into account in any detail by previous workers. The results of studies of this seawater inhibition are also reported.

Materials and Methods

Seawater samples used in these studies were collected during a cruise between Kodiak, Alaska, and Honolulu, Hawaii, in August–September, 1964. Waters were obtained from various depths in 3-liter Van Dorn samplers which had been sterilized previously with methanol. Each sample was immediately passed through a sterile HA Millipore filter, and the filtrate transferred to a sterile polypropylene bottle and frozen until bioassayed in the shore-base laboratory.

All bioassay glassware was washed with detergent, treated with chromic acid for 12–24 hours, rinsed thoroughly with deionized water, and baked for 12–24 hours at approximately 260 °C.

The composition of the basal medium is listed in Table I. In bioassays, the basal medium was prepared with the seawater sample. Dilutions of the samples were made with basal medium prepared in charcoal-treated aged seawater. Charcoal treatment was performed as described by Ryther and Guillard (8).

The samples to be bioassayed and the charcoal-treated seawater were sterilized by passage through PH Millipore filters. Fifty-milliliter quantities of each sample and 600 ml of the charcoal-treated seawater were supplemented with constituents of the basal medium which had been sterilized by passage through Morton fritted-glass filters, UF grade. Duplicate 5, 10, or 20 ml aliquots of each sample were transferred to 50-ml Erlenmeyer assay flasks and, where necessary, the volume brought to 20 ml with the basal medium prepared with charcoal-treated seawater. In all 20-ml samples (diluted or undiluted) the final concentrations of the basal medium constituents were the same as those listed in Table I. To one of each sample duplicate 1.0 $\mu\mu$g B_{12}

TABLE I

Composition of the basal medium

Constituents*	Amount, mg/liter
KNO_3	50.0
KH_2PO_4	3.50
Chelated metals	
$\quad CoCl_2 \cdot 6H_2O$	0.004
$\quad CuSO_4 \cdot 5H_2O$	0.004
$\quad FeCl_3 \cdot 6H_2O$	1.0
$\quad ZnSO_4 \cdot 7H_2O$	0.30
$\quad MnSO_4 \cdot H_2O$	0.60
$\quad Na_2MoO_4 \cdot 2H_2O$	0.15
$\quad EDTA$	6.0
$\quad Na_2SiO_3 \cdot 5H_2O$ (acidified to pH 8)	65.0
Vitamins	
$\quad$ Thiamine HCl	0.50
$\quad$ Nicotinic acid	0.10
$\quad$ Ca pantothenate	0.10
$\quad$ Inositol	5.0
$\quad$ Thymine	3.0
$\quad$ p-Aminobenzoic acid	0.010
$\quad$ Biotin	0.001
$\quad$ Folic acid	0.002

*Added to seawater.

43

per ml was added as an internal standard for determining the degree of inhibition (see below) of the particular seawater to the alga. A series of assay flasks containing 20 ml of the basal medium prepared with charcoal-treated seawater and 0–3.0 $\mu\mu$g B_{12} per ml served as external standards.

Cyclotella nana (clone 13-1) culture transfers were made every 3–4 days in 50 ml of enriched medium (basal medium prepared with charcoal-treated seawater + 100 μg B_{12} per liter) in 125-ml Erlenmeyer flasks. The inoculum for bioassay was prepared by transferring actively growing cultures to the basal medium. After 3–4 days, while the culture was in log phase, it was again inoculated into basal medium. The inoculum was ready to be added to bioassay samples after a further 3 days. Carry-over of B_{12} was then at a minimum but cells were still physiologically active. A subsequent inoculation into basal medium resulted in death of the organism. The correct preparation of this inoculum was a critical factor for the success of the method. Enough cells to give approximately 1×10^4 per ml were added to each bioassay flask; cell titers were determined with a hemocytometer.

Each flask was plugged with cotton wool enclosed in cheesecloth. Both the neck of the flask and plug were covered with parafilm to prevent evaporation of the samples during incubation. In more recent bioassays 50-ml micro-Fernbach flasks with DeLong style necks accommodating Morton stainless steel enclosures were used. All bioassay flasks were incubated at 22 $\pm$ 2 °C by placing the flasks over daylight fluorescent lights at an intensity of about 0.05 langleys per minute. After 47 hours, 1 ml of a solution containing 1 microcurie of ^{14}C as $Na_2^{14}CO_3$ in 5% w/v NaCl was added and incubation continued for 2 hours. For convenience, this is referred to as a 48-hour incubation. The cells were collected on an HA Millipore filter (1 in. diameter) and rinsed with filtered seawater. After the cells were dried, the ^{14}C in them was determined by a thin end-window Geiger counter.

From the curve of a plot of ^{14}C assimilation rate (counts per minute (c.p.m.) per hour of exposure to ^{14}C) vs. B_{12} concentrations in the external standards, the apparent concentration of this vitamin in the sample was obtained. Again using the external standard calibration curve the response of the added B_{12} in the internal standard in each sample was found. The fractional recovery was calculated and assumed to apply to the particular sample being assayed, and hence the correct B_{12} concentration could be estimated. If the sample was diluted this factor also had to be considered. The fractional amount of the internal standard which was not recovered was used to calculate the sample inhibition, i.e., if 0.2 of the added 1.0 $\mu\mu$g B_{12} per ml was not recovered, the sample had an inhibition value of 20%.

Results

The range of linearity of curves obtained from the external standards when B_{12} concentrations were plotted against c.p.m. per hour exposure to ^{14}C was dependent upon incubation time (Fig. 1). In addition to the 48-hour incubation series, one external standard series received ^{14}C after 22 hours and incubation was continued to 26 hours. ^{14}C was added to a third series after 71 hours and exposure continued for 2 hours. For convenience these last two times

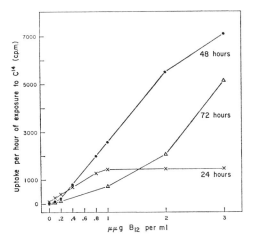

Fig. 1. Incorporation of ¹⁴C during 2- or 4-hour exposure to the isotope by cells of
C. nana grown for various times in different concentrations of B_{12}.

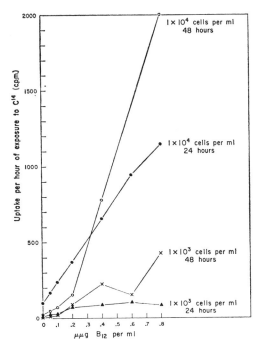

Fig. 2. Incorporation of ¹⁴C during 2- or 4-hour exposure to the isotope by cells of
C. nana using two inoculum concentrations in different levels of B_{12} after 24- and 48-hour
incubation.

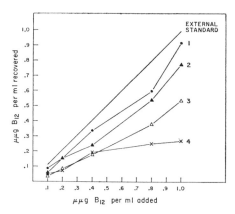

FIG. 3. Recovery of B_{12} added to 4 seawater samples giving different degrees of inhibition to the growth of *C. nana*. Curve for external standard assumes 100% recovery.

are referred to as 24- and 72-hour incubations, respectively. There was little difference in the ^{14}C uptake rates after 24 and 48 hours when less that 0.4 $\mu\mu g$ B_{12} per ml was present. With higher B_{12} concentrations uptake was greatest after 48 hours. A 72-hour incubation period appeared to be too lengthy.

The 48-hour curves have two slopes, one pertaining to the range 0–0.2 and the other to the range 0.2–2.0 $\mu\mu g$ B_{12} per milliliter. Since a number of points were used to determine the linear portions of the curve, the upper limit of the concentration range of the bioassay can be given with confidence as 3.0 $\mu\mu g$ B_{12} per milliliter. In a number of experiments it has been extended to 4.0 $\mu\mu g$ B_{12} per milliliter.

The influence of the *C. nana* inoculum titer on ^{14}C uptake in external standards after 24- and 48-hour incubations was investigated in another experiment. Initial inocula were 1×10^3 and 1×10^4 cells per milliliter. Figure 2 shows that ^{14}C incorporation was greater for the higher inoculum after 24 as well as after 48 hours. After 24 hours the higher inoculum gave linear uptake between 0 and 0.8 $\mu\mu g$ B_{12} per ml and the 48-hour result showed two linear portions on the curve, 0–0.2 and 0.2–0.8 $\mu\mu g$ B_{12} per milliliter. Inocula greater than 1×10^4 cells per ml gave too high a counting rate after 48 hours for good coincidence corrections to be possible with the counter after a 2-hour exposure to the isotope and it seemed undesirable to reduce this time very greatly.

The waters collected from the North Pacific Ocean were tested to see if they were inhibitory to *C. nana*, and if so, whether different levels of added B_{12} would affect the degree of inhibition. To a series of flasks containing a 1:2 dilution (with supplemented charcoal-treated seawater) of seawater, B_{12} was added in levels similar to those in the external standard series. The curves of Fig. 3 show the amounts of B_{12} recovered after 48 hours from the seawater samples, each of which differed in its inhibitory effect on *C. nana*. The B_{12} originally present in the water has been subtracted from the total amount detected. The amount of B_{12} recovered in each sample at each concentration was always less than that recovered for the same concentration in the charcoal-treated control water. The more toxic waters had greater inhibitory effects at the higher concentrations of vitamin B_{12}.

In several experiments waters collected from various locations at 60 and 120 m were used. Dilutions of $1:2$ and $1:4$ of the samples were prepared with a charcoal-treated, non-inhibitory water. Internal standards of 1.0 $\mu\mu$g B_{12} per ml were included. The data of Table II show that water from 60 m in station 23 was non-inhibitory and that the concentrations of B_{12} calculated from assays made at all dilutions were the same. Waters of 60 and 120 m in station 27 were also non-inhibitory. In station 24 at 120 m and station 28 at 60 and 120 m inhibition to *C. nana* was observed, but when this effect was taken into consideration, the concentration of B_{12} for each sample from each station was similar.

Seawater samples collected in a clean plastic bucket from the pier at the Scripps Institution were bioassayed. The undiluted seawater sample was 90% inhibitory to *C. nana* as calculated from the recovery of added B_{12} in the internal standard. From this value for percentage inhibition one could calculate an apparent concentration of B_{12} in the water of 10 $\mu\mu$g B_{12} per milliliter. Diluting the sample $1:2$ with charcoal-treated seawater resulted in 50% inhibition and an apparent concentration of 4.0 $\mu\mu$g B_{12} per milliliter. This was undoubtedly a more precise value than that obtained with the very inhibitory undiluted sample and probably a greater dilution to lessen even further the degree of inhibition would be desirable.

In the bioassay of a sample of seawater which was only slightly inhibitory, after an equal volume of charcoal-treated water had first been added, the standard deviation was found to be about 0.02 $\mu\mu$g B_{12} per ml on a concentration of 0.2 $\mu\mu$g per ml of B_{12}. Assuming this value, for the standard deviation to hold for even smaller concentrations of vitamin, one would judge the effective limit of detection of the method to be about 0.05 $\mu\mu$g B_{12} per milliliter. In a more inhibitory water having a concentration of 1.39 $\mu\mu$g B_{12} per ml, the standard deviation was found to be 0.27 $\mu\mu$g B_{12} per milliliter.

Discussion

C. nana in seawater containing adequate B_{12} undergoes about 2.5 generations a day. It was found that a terminal concentration of 2.2×10^5 cells per ml developed when 1.0 $\mu\mu$g B_{12} per ml was present. The average cell volume was 170 μ^3 and thus the number of molecules of B_{12} finally present in 1 μ^3 of cell was about 12. This confirms the results of Guillard and Cassie (3), who reported a value of 9 for *C. nana* having an average cell volume of 160 μ^3. In the external standard curve obtained after a 24-hour incubation, the uptake rates with between 1.0 and 3.0 $\mu\mu$g B_{12} per ml differed only slightly (Fig. 1) probably because the cells had similar responses to all B_{12} concentrations greater than 1.0 $\mu\mu$g B_{12} per milliliter. After 48 hours cells had responded more to the higher concentrations, and large differences in ^{14}C uptake were recorded. The less positive slope of the lower concentration portion of the curve (0–0.2 $\mu\mu$g B_{12} per ml) in the standard series after a 48-hour incubation probably resulted from medium B_{12} exhaustion.

Cell counts were made in several experiments and numbers were always proportional to B_{12} concentrations after sufficient growth time, i.e., 72 hours.

47

TABLE II

Percentage inhibition of response to added B_{12} together with the concentrations of B_{12} in seawater and its dilutions

| Station | Location | | Depth, m | Seawater | | | | | |
| | Long. °W. | Lat. °N. | | Undiluted | | Diluted 1:2 | | Diluted 1:4 | |
				% inhibition	B_{12} conc.,* $\mu\mu g/ml$	% inhibition	B_{12} conc.,* $\mu\mu g/ml$	% inhibition	B_{12} conc.,* $\mu\mu g/ml$
23	155 31.0	49 07.0	60	0	0.63	0	0.60	0	0.60
24	155 00	47 38.1	120	55	2.0	26	1.5	22	1.5
27	155 01.6	43 24.4	60	0	0.21	0	0.24		
			120	0	0.60	0	0.62		
28	155 09.0	42 09.2	60	36	0.23	26	0.28		
			120	39	0.44	0	0.42		

*Calculated from % recovery of added B_{12}, amount found in bioassay sample and dilution factor.

48

Cell division rates lagged behind the rates of ^{14}C incorporation, which appear to depend on cell vitality, and did not simply reflect cell biomass.

Gold (2) found a linear response to the vitamin by *C. nana* between 0.3 and 6.0 $\mu\mu$g B_{12} per ml after a 24-hour incubation. He obtained linearity to 0.1 $\mu\mu$g B_{12} per ml in other experiments. The results of the present studies indicate that the bioassay of freshly collected and non-charcoal-treated seawater after this time is not advisable. Concentrations of B_{12} are best determined after a 48-hour incubation during which time the algal cells have fully adjusted themselves to the new environment. When an internal standard is included, recovery of the added B_{12} is greater after 48 hours than after 24 hours. As an example, the water taken from one station at 30 m showed an apparent B_{12} concentration of 0.05 $\mu\mu$g per ml and 56% recovery in the internal standard after 24 hours, but after 48 hours these values were 0.1 $\mu\mu$g and 84%, respectively.

Gold (2) found different B_{12} concentrations after 24 hours in dilutions of the same water, and, since there was little difference in total cell numbers between treated and untreated seawaters, he concluded that the inhibitors affected only the lag phase. Inhibition by seawaters to *C. nana* and other algae has recently been reported by several workers (5, 9). Johnston (5) stressed the importance of antimetabolites or inhibitors in succession of algae in the sea. The action of inhibitors appeared to be lessened by the addition of water which is non-inhibitory, and the assumption that the charcoal-treated aged seawater used in these present studies was non-inhibitory seems justified.

In very inhibitory waters where amounts greater than 0.5 $\mu\mu$g B_{12} per ml may be present it is recommended that the samples be bioassayed after dilution with at least three times their volume of charcoal-treated water. An internal standard of 0.75 to 1.0 $\mu\mu$g B_{12} per ml is suggested for each sample. With larger amounts the inhibition appears to depend upon the vitamin concentration. With smaller additions the precision of the method will suffer. In all cases, however, internal standardization should be used to determine the percentage inhibition as this cannot be assumed to be negligible. If the inhibition much exceeds 30% the general precision of the assay will undoubtedly suffer.

Filter sterilization of seawater as well as of supplements is recommended In addition to removing particulate matter, the system is not exposed to heat which may cause a number of physicochemical changes. Autoclaving B_{12} stock solutions and standards, as a number of investigators have done, destroys B_{12} at pH 8 or above (8), and, although this can be accounted for by treating all samples and standards in a precisely similar manner, the procedure has little to recommend it.

C. nana responds to a number of B_{12} analogues. Activities of each of the analogues expressed in relation to 100 for crystalline vitamin B_{12} were as follows: pseudovitamin B_{12}, 60; factor A, 72; factor B, 65; factor III, 78; benzimidazole cobalamine, 43; 2-mercaptoadenine cobalamine, 79; 5-methylbenzimidazole cobalamine, 47. Etiocobalamine phosphoribose showed no activity at the concentrations tested. The concentration of B_{12} in samples of seawater in which *C. nana* is the bioassay organism can, therefore, reflect the presence of any of the different analogues of the vitamin.

49

A large number of samples from various locations in the North Pacific Ocean are currently being bioassayed using the technique described in this paper, and the results for B_{12} concentrations and seawater inhibition will be reported in a later communication.

Acknowledgments

We thank Dr. J. D. H. Strickland for constructive criticism in the preparation of this manuscript, and Dr. L. Provasoli for supplying the vitamin B_{12} analogues. This work was supported in part by the Marine Life Research Program, Scripps Institution of Oceanography's component of the California Cooperative Oceanic Fisheries Investigations, a project sponsored by the Marine Research Committee of the State of California, and in part by the United States Atomic Energy Commission, Contract AT(11-1)-34, Project 108.

References

1. BELSER, W. L. 1963. Bioassay of trace substances. *In* The sea. *Edited by* M. N. Hill. Vol. 2. Interscience, N.Y. pp. 220–231.
2. GOLD, K. 1964. A microbiological assay for vitamin B_{12} in seawater using radiocarbon. Limnol. Oceanog. **9**, 343–347.
3. GUILLARD, R. R. L. and CASSIE, V. 1963. Minimum cyanocobalamin requirements of some marine centric diatoms. Limnol. Oceanog. **8**, 161–165.
4. HUTNER, S. H. and PROVASOLI, L. 1964. Nutrition of algae. Ann. Rev. Plant Physiol. **15**, 37–56.
5. JOHNSTON, R. 1963. Seawater, the natural medium of phytoplankton. I. General features. J. Marine Biol. Assoc. U.K. **43**, 427–456.
6. MENZEL, D. W. and SPAETH, J. P. 1962. Occurrence of vitamin B_{12} in the Sargasso Sea. Limnol. Oceanog. **7**, 151–154.
7. PROVASOLI, L. 1963. Organic regulation of phytoplankton fertility. *In* The sea. *Edited by* M. N. Hill. Vol. 2. Interscience, N.Y. pp. 165–219.
8. RYTHER, J. H. and GUILLARD, R. R. L. 1962. Studies of marine planktonic diatoms. II. Use of *Cyclotella nana* Hustedt for assays of vitamin B_{12} in seawater. Can. J. Microbiol. **8**, 437–445.
9. SMAYDA, T. J. 1964. Enrichment experiments using marine centric diatom *Cyclotella nana* (clone 13-1) as an assay organism. Proc. Symp., Exptl. Marine Ecol. Graduate School Oceanog., Univ. Rhode Island, **2**, 25–32.

Biochemical Analysis of Preserved Zooplankton

H. Fudge

Biochemical analysis of zooplankton at sea presents many difficulties so specimens caught during a voyage are preserved until analysis can be performed in a land-based laboratory. The literature[1-3] indicates that two methods of preservation are generally used—deep-freezing[3,4] and preservation in formalin[1,4]. Most analyses have been carried out on material preserved thus, with no direct comparison with fresh material, and so there must be some doubt as to the accuracy of the values obtained. I have investigated several different methods of preservation in order to assess their suitability for use before chemical analysis of zooplankton.

The brackish-water mysid *Neomysis integer* was used because it was locally abundant and has a well known biochemical composition[5]. Five methods of preservation were used: 10 per cent v/v formalin; 10 per cent tri-chloracetic acid (TCA); 70 per cent ethanol; deep-freezing at −25° C, and freeze-drying. Specimens of *Neomysis* were preserved by each method for a period of 3–5 weeks, and additional specimens were preserved in formalin for 24 h.

Preserved and fresh *Neomysis* were analysed for dry matter, ash, chitin, lipid and carbohydrate using the methods of Raymont et al.[5]. Protein was measured by the biuret method[5].

For each batch, usually five determinations were made for dry matter, ash and chitin, and ten determinations for lipid, protein and carbohydrate were made. In addition to oven dry weights, a supplementary dry weight was determined by weighing before, and after, the freeze-drying process. The results shown in Table 1 are expressed as percentage dry weight, except the dry weight data which are given as percentage wet weight.

Clearly there are some very marked differences between the fractions of fresh and preserved *Neomysis*. In view of the limited number of determinations made, however, only gross changes can be discussed.

The results for the fresh material are in fair agreement with those quoted by Raymont et al.[5], and this suggests

Table 1. MEAN VALUES AND TOTALS OF ALL FRACTIONS OBTAINED FOR EACH METHOD OF PRESERVATION

Sample and preservation time	Dry weight	Ash	Chitin	Lipid	Protein	Carbo-hydrate	Total
Fresh	22 ± 1	12 ± 0	3 ± 0	11 ± 3	73 ± 7	3 ± 0	103%
24 h forma-lin	21 ± 0	12 ± 0	3 ± 0	8 ± 1	6 ± 3	3 ± 0	32%
4 weeks for-malin	20 ± 1	9 ± 2	3 ± 1	15 ± 2	5 ± 5	9 ± 6	41%
3 weeks TCA	36 ± 2	1 ± 0	2 ± 0	7 ± 2	44 ± 4	1 ± 0	55%
4 weeks ethanol	17 ± 2	10 ± 3	5 ± 1	9 ± 1	92 ± 11	2 ± 0	118%
5 weeks deep freeze	29 ± 2	11 ± 0	5 ± 1	15 ± 3	79 ± 7	3 ± 2	113%
4 weeks freeze dried	20 ± 1*	13 ± 0	4 ± 0	14 ± 2	74 ± 9	2 ± 0	107%

* The freeze dried material was, of course, anhydrous, the dry weight quoted here is the figure obtained for drying fresh material by freeze drying.

that the analyses were performed satisfactorily.

Material preserved for 24 h in formalin showed a large decrease in the protein fraction, and further preservation (4 weeks) gave an increase in the carbohydrate reaction. Thus formalin gives useful results with a short time of preservation, except for the protein fraction (estimated by the biuret reaction), but this problem might be solved by a Kjeldahl determination of nitrogen followed by conversion to protein ($N \times 6 \cdot 25$).

With the exception of dry matter content, which rose above that of fresh tissue, all fractions from material preserved in TCA decreased, notably ash, protein and carbohydrate. With animals preserved in alcohol, proteins gave a much higher value than the fresh tissue protein, and only the carbohydrate fraction can be regarded as a reliable result. Deep-freezing showed a small increase in most fractions, thus producing a total of 113 per cent. The dry weight value was higher than the value for fresh tissue because of dehydration during refrigeration. Freeze drying gave values for all fractions which were close to those obtained from fresh specimens. Lipid is a very variable fraction even in fresh animals and any variation in preserved material is only to be expected.

As far as biochemical analyses are concerned, a good method of tissue preservation should produce as little alteration as possible in the chemical nature of the sample. Thus, formalin, TCA and alcohol are not suitable except for one or two fractions, although formalin is satisfactory over a short period for most fractions. Freeze drying clearly offers a reliable method, and deep freezing is also reliable provided that dry weight determinations are made simultaneously with analyses of other fractions.

[1] Bogorov, B. G., J. Mar. Biol. Assoc. UK, 19, 585 (1934).
[2] Gillam, A. E., El Ridi, M. S., and Wimpenny, R. S., J. Exp. Biol., 16, 71 (1939).
[3] Raymont, J. E. G., Austin, J., and Linford, E., Deep Sea Res., 14, 113 (1967).
[4] Vinogradova, Z. A., Dokl. Akad. Nauk SSSR, 133 (3), 680 (1960).
[5] Raymont, J. E. G., Austin, J., and Linford, E., J. Cons. Intern. Explor. Mer., 28, 354 (1964).

Polonium-210 and lead-210 in the marine environment

L. V. Shannon

R. D. Cherry

and

M. J. Orren

The radioactivity resulting from lead-210 and polonium-210 in the natural environment has been the subject of increasing attention in recent years. Radon-222 emanating from the surface of the land escapes into the atmosphere, where it decays via short lived daughters to lead-210 and its daughters. Several authors (Burton and Stewart, 1960; Blifford et al., 1952; Paterson and Lockhart, 1964; Lambert et al., 1966; Pierson et al., 1966; Ter Haar et al., 1967), have given data for lead-210 and polonium-210 in the atmosphere and rains. The lead-210 activity of the lower troposphere is about 1×10^{-2} d.p.m./kg while the ratio of the activities of polonium-210 to lead-210 in ground level air is typically about 0·14 (Burton and Stewart, 1960; Pierson et al., 1966). Data on these nuclides in rain over land are numerous for localities in the northern hemisphere (Burton and Stewart, 1960; Blifford et al., 1952; Pierson et al., 1966; Ter Haar et al., 1967) but few results are available for the southern hemisphere (Blifford et al., 1952; Pierson et al., 1966). From these publications a typical value for lead-210 in rains over the northern hemisphere lies in the range 1–5 pc/l and somewhat lower (about 0·5 to 2 pc/l) in the southern hemisphere. The polonium-210 to lead-210 ratio in rain water is about 0·1; i.e. similar to that observed for the troposphere (Burton and Stewart, 1960; Lambert and Nezami, 1965). Calculations of the residence time of lead-210 in the atmosphere based on measurements of the polonium-210/lead-210 ratio have been made (BURTON and STEWART, 1960; PIERSON et al., 1966; LAMBERT and NEZAMI, 1965) and provide values ranging from 20 to 40 days.

Several authors have reported lead-210 and polonium-210 in biological materials of terrestrial origin, including foodstuffs, human tissues, blood, tobacco and various animals (HILL, 1965; BEASLEY and PALMER, 1966; LITTLE and McGANDY, 1966 HOLTZMAN, 1966) while more recently data have been provided for the concentration of these nuclides in marine life. Unsupported polonium-210 has been reported in marine plankton (CHERRY, 1964; SHANNON and CHERRY, 1967) at levels of about 4·7 pc/g dry weight for zooplankton and 2·8 pc/g dry weight for phytoplankton. Lead-210 and polonium-210 measurements on several marine biological materials from the Atlantic and Pacific Oceans have also been made (HOLTZMAN, 1969; BEASLEY et al., 1969).

Very little work appears to have been done on either polonium-210 or lead-210 in sea water. Lead-210 levels were first reported in 1961 by RAMA et al. (1961) and subsequently considered in more detail by GOLDBERG (1963). These observations were made on large sea water samples collected in the Eastern Pacific. The method employed by RAMA et al. (1961) involved coprecipitation of the lead followed by anion exchange and beta-counting. The lead-210 content ranged from about 0·10 d.p.m./l in the upper water layers to 0·28 d.p.m./l. at a depth of 2000 m. From these results, after making various assumptions, a biological removal time for lead-210 of 2 years was calculated.

To the best of our knowledge no values for the polonium-210 level in sea water have appeared in scientific journals, although two institutional reports (FOLSOM, 1966; REPORT BATTELLE–NORTHWEST, 1968) have provided some data. The method of Folsom used a scavenger (ferric hydroxide) in fifty-litre sea water samples followed by alpha-analysis. A mean value of 0·037 pc/l (or 0·08 d.p.m./l) was reported for samples collected from Scripps Pier.

In the present paper we shall give details of a simple method for determining polonium-210 and lead-210 in sea water and we shall present data on the distribution of these nuclides in sea water and plankton off South Africa during March 1969.

METHODOLOGY

Sea water

From theoretical considerations it was evident that the polonium-210 content of sea water would be extremely small and that we would be dealing with a micro-micro trace element (viz. at the level of 1 part in 10^{20} or 10^{21}). Polonium-210 is strongly adsorbed onto glass (FLYNN, 1968) and this necessitates a cautious sampling and storing procedure. Only a brief outline of the technique is given as full details are to be published elsewhere.

One and a half-litre sea water samples were collected using a plastic sampler, acidified with hydrochloric acid and stored frozen. Prior to analysis the samples were thawed and the pH adjusted to 2. The polonium was extracted as the ammonium pyrrolidine dithiocarbamate chelate into methyl isobutyl ketone. The organic solution was subsequently evaporated to near dryness and the residue taken up in hydrochloric acid. The polonium-210 was then plated out using a teflon deposition cell and the technique of FLYNN (1968). After plating the solution was stored for two months and then replated. By allowing any lead-supported polonium-210 to build up, an estimate of the lead content was made (HILL, 1965; HOLTZMAN, 1969; BEASLEY et al., 1969). Samples were counted using a zinc sulphide sealed sample phosphor system and conventional photomultiplier assemblies. The characterisation of the activity of all samples was checked on a multi-channel semi-conductor detector system.

A blank count rate of 1·00 ± 0·05 counts per hour was obtained using the same reagents and technique on doubly distilled water. Calibration on the entire extraction and plating technique was done using lead–polonium-210 (in equilibrium) spikes ranging from 0·04 pc/l to 600 pc/l. Overall efficiencies were 92 % for polonium-210 and 85 % for lead-210. All activities are expressed as at time of collection, and include corrections for unsupported polonium-210 decay and supported polonium-210 build-up.

Plankton

Plankton samples were collected and dried as described by SHANNON (1968). A portion of each sample was wet ashed in nitric and perchloric acids (BLACK, 1961; HOLTZMAN, 1963) and the polonium plated out using the technique of FLYNN (1968). Replating after two months provided an estimation of the lead-210 contribution. Total alpha-counting on the other portion of the dried plankton sample (CHERRY, 1964; TURNER *et al.*, 1958; SHANNON and CHERRY 1967) provided a check on the polonium-210 measurements.

RESULTS AND DISCUSSION

Sea water samples were collected at a standard depth of 20 m at stations on separate cruises by R.S. *Africana II* and R.S. *Sardinops* during March 1969. This depth was chosen as representative of the upper mixed layer and at the same time free of possible contamination from the ship. The results of the polonium-210 and lead-210 measurements are given in Table 1. Activities have been expressed in 10^{-15} c/l (1 × 10^{-15} c/l is equivalent to 0·0022 d.p.m./l).

Table 1. Polonium-210 and lead-210 in sea water at 20 m, March 1969

Sample No.	Ships Station No.	Date	Latitude	Longitude	Polonium-210 (10^{-15} c/l)	Lead-210 (10^{-15} c/l)
Cruise A: R. S. Sardinops						
S1	F5990	18/3/69	33°51'S	18°17'E	15 ± 2	30 ± 9
S2	F5991	18/3/69	34°16'S	17°40'E	8 ± 2	17 ± 8
S3	F5992	18/3/69	34°42'S	17°00'E	8 ± 2	<10
S4	F5993	18/3/69	35°07'S	16°23'E	11 ± 2	31 ± 9
S5	F5994	19/3/69	34°12'S	15°47'E	17 ± 2	35 ± 9
S6	F5995	19/3/69	32°40'S	14°33'E	15 ± 5	16 ± 7
S7	F5996	20/3/69	32°40'S	15°44'E	18 ± 2	17 ± 5
S8	F5997	20/3/69	32°45'S	16°33'E	15 ± 2	<10
S9	F5998	20/3/69	32°42'S	17°18'E	30 ± 5	15 ± 5
S10	F5999	20/3/69	32°40'S	17°53'E	21 ± 2	<10
S11	F6000	21/3/69	33°25'S	17°59'E	28 ± 2	<10
Cruise B: R.S. Africana II						
S12	A5127	6/3/69	34°33'S	21°05'E	11 ± 2	32 ± 9
S13	A5131	6/3/69	36°30'S	21°36'E	33 ± 2	37 ± 9
S14	A5133	7/3/69	38°27'S	22°00'E	30 ± 3	55 ± 9
S15	A5135	8/3/69	40°26'S	22°24'E	27 ± 2	51 ± 9
S16	A5148	13/3/69	39°10'S	17°04'E	29 ± 3	38 ± 9
S17	A5150	14/3/69	37°45'S	17°25'E	28 ± 3	19 ± 10
S18	A5152	15/3/69	35°48'S	18°53'E	12 ± 2	24 ± 10
S19	A5164	17/3/69	36°59'S	13°45'E	13 ± 2	84 ± 15
S20	A5166	18/3/69	35°45'S	15°32'E	41 ± 4	<10
S21	A5168	19/3/69	34°53'S	17°03'E	19 ± 2	37 ± 12
S22	A5171	19/3/69	33°51'S	18°17'E	24 ± 2	91 ± 18
S23	A5179	23/3/69	35°17'S	12°25'E	14 ± 2	74 ± 11
S24	A5180	24/3/69	32°45'S	12°15'E	15 ± 2	135 ± 16
S25	A5181	24/3/69	32°35'S	13°17'E	14 ± 2	98 ± 16
S26	A5182	24/3/69	32°42'S	14°32'E	15 ± 2	11 ± 9
S27	A5183	24/3/69	32°43'S	15°45'E	16 ± 2	47 ± 9
S28	A5184	25/3/69	32°41'S	16°37'E	28 ± 3	68 ± 15
S29	A5187	25/3/69	32°40'S	17°54'E	18 ± 2	15 ± 10

The stations occupied during cruise A (R.S. *Sardinops*) were situated along the West Coast close to Cape Town. Corresponding plankton samples were collected at these stations for biological removal estimates.

Cruise B (R.S. *Africana II*) covered a far wider area (this cruise in fact formed part of a combined survey of the Agulhas Current). The sea water samples for polonium and lead analysis were collected at 18 stations selected to provide a good coverage. Unfortunately, owing to the tight schedule on this cruise, corresponding plankton samples were not collected at any of the stations.

The horizontal distributions of polonium-210 and salinity at 20 m (cruise B only) are shown in Fig. 1. For the purposes of comparison, the salinity distribution has been drawn using data only from these stations at which polonium-210 measurements were made. The basic trend remains unaltered by the inclusion of the additional available values.

Polonium-210 and lead-210 activities in plankton from Cruise A are tabulated in Table 2.

The polonium-210 content of the surface layer of the sea around the Cape of Good Hope during March 1969 varied between 8×10^{-15} c/l and 41×10^{-15} c/l with

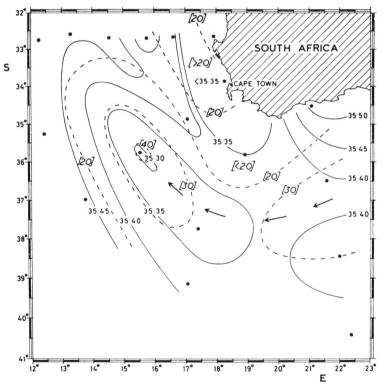

Fig. 1. Comparison of salinity (‰—solid lines) and polonium distribution (10^{-15} c/l—values in brackets, broken lines), Cruise B, March 1969.

Table 2. Polonium-210 and lead-210 in plankton, cruise A, March 1969

Sample No.	Ship Station No.	Depth (m)	Polonium-210 pc/kg (wet)	Lead-210 pc/kg (wet)
Zooplankton				
Z1	F5990	70–0	258 ± 6	23 ± 2
Z2	F5991	100–0	368 ± 4	17 ± 1
Z3	F5992	100–0	395 ± 11	40 ± 3
Z4	F5993	100–0	600 ± 11	62 ± 4
Z5	F5994	100–0	697 ± 13	62 ± 4
Z6	F5995	100–0	291 ± 12	30 ± 2
Z7	F5996	100–0	570 ± 10	51 ± 3
Z8	F5997	100–0	433 ± 8	21 ± 2
Z9	F5998	100–0	272 ± 7	27 ± 2
Z10	F5999	40–0	226 ± 6	9 ± 1
Z11	F6000	80–0	276 ± 7	19 ± 2
Phytoplankton				
P9	F5998	50–0	165 ± 3	69 ± 4
P10	F5999	40–0	42 ± 2	12 ± 1
P11	F6000	50–0	109 ± 4	20 ± 2

a mean value of 20×10^{-15} c/1 or 0·044 d.p.m./l. During January and February, while the experimental technique was being tested, 4 samples were collected off rocky beaches around the Cape Peninsula. The data have not been included in Table 1 as it is considered that the proximity of rocks, sand and effluent outlets may have introduced contamination, but nevertheless the mean polonium-210 activity of these samples was 30×10^{-15} c/1. This figure is in good agreement with FOLSOM (1966) who obtained values of 30×10^{-15} and 37×10^{-15} c/1 for samples collected off the Scripps Pier.

The lead-210 values presented in Table 1 are subject to large errors due to counting statistics. A conservative estimate of the effective lower detection limit of the method for lead-210 (using 1·5-l. samples and two month storage before counting) is about 10×10^{-15} c/1 and results lower than this value have simply been tabulated as $<10 \times 10^{-15}$ c/1. The lead-210 activity of the surface layer during March 1969 ranged from "zero" (i.e. $<10 \times 10^{-15}$ c/1) to 135×10^{-15} c/1 with a mean value of 38×10^{-15} c/1, or 0·084 d.p.m./l. RAMA et al. (1961) obtained a value of about 0·11 d.p.m./l for the upper water layer in the Pacific, a figure in good agreement with our data.

It is evident from the data that there are considerable variations in the lead-210 and polonium-210 content. This is to be expected when the modes of influx and removal of these nuclides are considered. Attempts were made to correlate the polonium-210 content with currents and water movements around the Cape, and inspection of Fig. 1 shows a marked similarity between the polonium-210 and salinity distribution. Arrows have been drawn in the figure to show the movement of a component of the Agulhas Current around the Cape. The salinity of typical Agulhas Water is 35·2–35·4‰ or about 0·3‰ lower than South Atlantic Subtropical Surface Water (SHANNON, 1970). Thus the penetration of a tongue of Agulhas Water into the West Coast region about 200 miles off shore as well as an eddy in this intrusion (centred at 35°45'S, 15°32'E) can be seen in Fig. 1. The polonium-210 data (Fig. 1) show this intrusion and eddy very clearly. Typically, polonium-210 values in the intrusion were greater than 25×10^{-15} c/1 when the salinity was lower than 35·4‰.

5

The eddy had a salinity of 35·29‰ and a polonium-210 content of 41 × 10^{-15} c/l. Although data are limited the polonium-210 distribution does appear to be closely related to water movements in the off shore region. Closer inshore, upwelling and high biological activity makes the interpretation of the data difficult. The statistical errors of the lead-210 results precluded any accurate comparison with water movements, but it does appear that on the average Agulhas Water had a lower lead-210 content than the surrounding water masses.

A single depth profile at 32°40′S, 14°33′E comprising 4 samples at intervals between the surface and 600 m showed lead-210 and polonium-210 values little different from the mean 20 m values: no indication of variations with depth could be established.

The lead-210/polonium-210 activity ratio in sea water was about 2 on the average. Individual samples, however, showed an interesting trend. In most samples the activity ratio was greater than unity, but in about one third of the samples the ratio was near to, or less than one. Of these samples, 7 were collected in an area lying north–west of Cape Town (bounded by the coast and the 14°E meridian) and two were from the eddy of Agulhas Water mentioned earlier. The possibility exists of using the lead-210/polonium-210 ratio for estimating the age of oceanic waters, but this can only be done when the complex processes involved are better understood.

At stations on cruise A plankton samples were collected. The plankton sample numbers correspond to sea water sample numbers but have a prefix Z in the case of zooplankton and P in the case of phytoplankton. Attempts were made to establish a relationship between the activity of plankton per litre of sea water and the activity of the sea water, but no such trend was apparent. Plankton data have been tabulated in activity per wet weight of plankton rather than in activity per litre of sea water. In all cases the contribution of the netted plankton to the total polonium-210 activity of the sea water was less than 0·5%. The mean lead-210 and polonium-210 activities of the zooplankton samples were 33 pc/kg (wet) and 399 pc/kg (wet) respectively. For the 11 zooplankton samples the average lead-210/polonium-210 activity ratio was 0·083 and the positive correlation coefficient of 0·89 between lead-210 and polonium-210 was statistically highly significant. The corresponding atom ratio of lead-210/polonium-210 is about 5. For phytoplankton the lead-210/polonium-210 activity ratio appears to be in the range 0·2–0·4 with an atom ratio of about 18.

The lead-210 content of both zooplankton and phytoplankton was similar, namely about 33 pc/kg wet material. The polonium-210 content of zooplankton (399 pc/kg wet material) was however, a factor of about four higher than that in phytoplankton. This factor confirms earlier data (SHANNON and CHERRY, 1967; SHANNON, 1969) which showed that polonium-210 accounted for most of the alpha-activity in zooplankton and a smaller fraction in phytoplankton. From the data presented the enrichment factors of lead-210 and polonium-210 from sea water to wet zooplankton are about 870 and 20,000 respectively. These results are in good agreement with our earlier less accurate estimates (SHANNON and CHERRY, 1967). For phytoplankton the enrichment factors are about 890 for lead-210 and 5400 for polonium-210.

In order to calculate the removal times for polonium-210 and lead-210 it is necessary to have an accurate estimate of the rate of influx of these nuclides into the sea. RAMA et al. (1961) quote the input into the atmosphere of radon-222 as 42 atoms per minute per square centimeter of land. This implies that the average standing crop of radon in the atmosphere is 42 disintegrations per minute per square centimeter of land area. In another calculation LAMBERT *et al.* (1966) estimated that $2 \cdot 5 \times 10^{25}$ radon atoms are emanated annually by the continents. Excluding Antarctica this yield corresponds to 35 atoms per minute per square centimeter of land area. The agreement with the radon value of RAMA *et al.* is reasonable. The land area of the southern hemisphere (excluding Antarctica which is covered by a layer of ice) is about $13 \cdot 5\%$ of the toal hemispherical area (SVERDRUP *et al.*, 1942). Assuming no hemispherical interchange of atmosphere on the short time scale involved, assuming equilibrium between radon emanated and lead-210 deposited (LAMBERT *et al.*, 1966) and assuming equal deposition of lead-210 over land and oceanic areas, then these values imply that on the average something like 5 atoms of lead are deposited per minute per square centimeter of total area in the southern hemisphere, i.e. about one third of the value for the northern hemisphere. This estimate should be considered as a maximum for deposition over oceanic areas because the short time scale will presumably imply a preferential deposition over land rather than oceanic areas.

Data extracted from PIERSON *et al.* (1966) indicates that during the period 1961–1964 the average lead-210 contents of coastal rain in the northern and southern hemispheres (excluding Ottawa and Melbourne which are considered as being continental) were $2 \cdot 46$ pc/l and $0 \cdot 88$ pc/l respectively. Thus on the average the southern hemispherical oceanic areas have a lead-210 content in rain which is about 36% of that of the northern hemisphere, i.e. a figure in fair agreement with the effective distribution of land in the two hemispheres. Very little data on true oceanic rainfall is available. Figures cited by LOEWE (1957) indicate a rainfall over the sea between latitudes 45°S and 70°S of about 80 cm per year. BLIFFORD *et al.* (1952) quote the total rainfall over land as 99×10^3 km^3/yr, which implies an average rainfall over land of 67 cm/year. If we assume that the average rainfall over the sea in the southern hemisphere is 70 cm/year and assume further that this rainfall contains $0 \cdot 88$ pc/l of lead-210, then each square centimeter of area receives $0 \cdot 062$ pc of lead-210 per year or $4 \cdot 2$ atoms per minute. This figure is to be compared with the "maximum" fall-out estimate of about 5 atoms/min per cm^2 calculated above from the consideration of land areas and production rates of radon-222. It is perhaps a little surprising that these values are in such close agreement.

It will therefore not be unreasonable if we use a southern hemisphere atmospheric fall-out rate of lead-210 of 5 atoms/min per cm^2. Suppose we consider a very simple two layer ocean with the atmospherically produced lead-210 fall-out distributed throughout the upper 100 m. The input of lead-210 into this layer from the atmosphere is therefore $0 \cdot 5$ atoms/min per litre. Let us also assume that no vertical mixing takes place. Thus the following balance equation may be set up

$$\frac{\mathrm{d}N}{\mathrm{d}t} = 0 = I - \lambda N - RN$$

where N is the concentration of lead-210 in the mixed layer, I is the total input rate of lead-210 into the mixed layer from both atmospheric lead-210 fall-out and decay of radium-226 in sea water, λ is the radioactive decay constant for lead-210, and R is the probability per unit time of a lead-210 atom being removed by biological and inorganic removal processes. R is called the "rate constant" while the reciprocal of R is customarily referred to as the "removal time" for the isotope concerned.

Two recent papers (BROECKER et al., 1967; SZABO, 1967) have given the radium-226 content of the surface water of the Atlantic and Pacific Oceans as 4×10^{-14} g/l. Broecker et al. have, however, shown that radon-222 is not in equilibrium with its parent in the surface layer, the radon-222/radium-226 activity ratio for the upper 100 m being between 0·4 and 0·8. For the purpose of the calculation we will take the radon-222 concentration in the surface layer as being equivalent to 3×10^{-14} g/l of radium-226. This implies a production rate of lead-210 of 0·067 atoms/min per litre, giving the total input $I = 0.5 + 0.067 = 0.57$ atoms of lead-210 per min per litre. Inserting this value and our measured mean lead-210 content of 0·084 d.p.m./l into the balance equation gives $R = 0.19$ yr^{-1}. The implied removal time for lead-210 is therefore about 5 yr. RAMA et al. (1961) found a removal time for lead-210 of 2 yr for the North Pacific Ocean. The difference between their estimate and ours is due largely to the fact that they used the lead-210 fall-out rate applicable in the Northern Hemisphere, viz. 15 instead of 5 atoms/cm^2 per min.

A similar balance equation can be set up for polonium-210. From rainfall data it appears that in rain the polonium-210/lead-210 activity ratio is about 0·1. This corresponds to an input from the atmosphere of less than 0·001 atoms of polonium-210 per min per litre in the mixed layer (i.e. a negligible amount). The input from lead-210 decay (based on our data) is 0·084 atoms/min per litre while the polonium content of the mixed layer was found to be 0·044 d.p.m./l. Inserting these values in the balance equation we find $R = 1.7$ yr^{-1} i.e. a removal time for polonium-210 of 0·6 yr.

It is interesting to note how the *atom* ratio of lead-210 to polonium-210 decreases steadily from the fall-out rainwater to the zooplankton. In rainwater this ratio is 581, in sea water 111, in phytoplankton 18 and in zooplankton 5. It is furthermore perhaps significant that our data plus the rate factors calculated above imply that the ratio of lead-210 to polonium-210 atoms removed per unit time from the sea water in the upper mixed layer in order to maintain balance is about 12. The similarity of this figure to the corresponding atom ratio in plankton might be an indication that *biological* removal by plankton is the prime removal mechanism for both polonium-210 and lead-210. Insufficient supplementary data is available to test the validity of this possibility but certain problems are immediately apparent. Consider, for example, the problem of the quantity of living organic matter in the upper mixed layer. The biomass of the mixed layer as determined by hauls of standard plankton nets on Cruise A was 1.5×10^{-4} g wet plankton per litre of sea water. It is difficult to ascribe removal to plankton alone from consideration of the above value. If we take the polonium-210 content of plankton as 400 pc/kg wet weight then in order to maintain the removal of polonium-210 from the sea water as indicated by our data, the plankton would have to be replaced approximately every 16 hr. The typical biomass figure quoted above is, however, almost certainly a low estimate on account of the selectivity of the standard plankton nets. Recent unpublished measurements on the particulate and carbon content of sea water made by the Division of Sea Fisheries here indicates that about 2 mg (dry) organic material is present in a litre of sea water. What percentage of this is living matter is uncertain, but nevertheless even if it is as low as 10% the resulting biomass estimate is higher than that from net data. A longer and more acceptable plankton replacement time would then result.

It is also possible that the removal of polonium-210 and lead-210 from the sea water may be associated with diurnal migrations of zooplankton. Vinogradov (1961) postulated a cellular scheme of migrations of zooplankton. He considers that the most important way of supplying food to the deep ocean is by active transport of organic matter from the upper zone by migrating animals. At night a great number of interzonal animals ascend to the surface to feed on phytoplankton and minute animals, descending at day to depths of some hundred metres. Thus organic matter produced at the surface is actively carried to the deep ocean along a "ladder of migrations." His hypothesis is largely based on the fact that apart from deep-bottom animals, the main mass of the deep sea pelagic animals are carnivorous and not detritus eaters. Such a cellular scheme could indeed be responsible for a relatively rapid system of removal of lead-210 and polonium-210 from the surface layer, but must be regarded as speculative until such time as more detailed experimental data are available.

A further possibility, viz. that the decreasing lead-210 to polonium-210 ratio from rain water to zooplankton could perhaps be explained by the supply of polonium-210 to the surface water from bottom water has been suggested to us by Dr. D. P. Kharkar. We have not investigated this possibility in any detail, but would re-iterate that the data from the single depth profile we have available did not show any indication of variation in the lead-210 or polonium-210 content over the 600 m range.

Conclusion and Summary

We feel that the data in this paper show clearly:

(a) That the levels of polonium-210 and lead-210 can conveniently and rapidly be measured in small sea water samples by solvent extraction followed by alpha counting and/or alpha-spectrometry. One and a half litres of sea water is adequate for polonium-210 measurements, but is insufficient for rapid and accurate lead-210 determinations. A five litre sample would be more satisfactory

(b) That the levels of polonium-210 and lead-210 in surface sea water around South Africa during March 1969 were about 20×10^{-15} c/l (0.044 d.p.m./l) and 38×10^{-15} c/l (0.084 d.p.m./l) respectively.

(c) That real variations in the lead-210 and polonium-210 content in the mixed layer exist and that these variations are, in part at least, due to differing currents and water masses. The possibility of using these isotopes as natural oceanographic tracers is apparent.

(d) That lead-210 and polonium-210 are not in equilibrium in the marine environment, the lead-210/polonium-210 activity ratios in sea water, phytoplankton and zooplankton being about 2, $\frac{1}{3}$ and $\frac{1}{12}$ respectively.

(e) That a correlation exists between the lead-210 and polonium-210 content of zooplankton, the activity of the former being approximately one twelfth that of the latter.

In more speculative vein, the data indicate:

(f) That the removal times for lead-210 and polonium-210 from the upper mixed layer are about 5 yr and 0.6 yr respectively. These times could be significantly altered

61

if a more complex oceanographic model is considered and also if the lead-210 fall-out rate over the ocean is different to the figure we have assumed.

(g) That the atom ratio of lead-210/polonium-210 removed from the upper 100 m mixed layer is of the same order of magnitude as the atom ratios of these nuclides in zooplankton and phytoplankton.

It is in any event clear that many intriguing problems relating to the lead-210 and polonium-210 biogeochemical balance in the marine environment exist. Before the balance can be determined with certainty, a good estimate of the fall out of lead-210 over the open ocean will have to be made. In addition measurements will have to be made on the polonium-210 and lead-210 content of particulate organic and inorganic material present in sea water.

Acknowledgements—We are very appreciative of the extensive facilities provided by the Division of Sea Fisheries and the Staff Research Fund, University of Cape Town. We must also acknowledge financial support from the CSIR Oceanographic Research Unit at the University of Cape Town.

We should like to thank Mr. P. E. WRIGHT for assisting with the chemical analyses and Mr. T. BLAMIRE for drawing the figure.

REFERENCES

BEASLEY T. M., OSTERBERG C. L. and JONES Y. M. (1969) Natural and artificial radionuclides in sea foods and marine protein concentrates. *Nature* **221**, 1207–1209.

BEASLEY T. M. and PALMER H. E. (1966) Lead-210 and polonium-210 in biological samples from Alaska. *Science* **152**, 1062–1064.

BLACK S. C. (1961) Low level polonium and radiolead analyses. *Health Phys.* **7**, 87–91.

BLIFFORD I. H., LOCKHART L. B., JR. and ROSENSTOCK H. B. (1952) On the natural radioactivity in the air. *J. Geophys. Res.* **57**, 499–509.

BROECKER W. S., HUI LI Y. and CROMWELL J. (1967) Radium-226 and radon-222: Concentration in Atlantic and Pacific Oceans. *Science* **158**, 1307–1310.

BURTON W. M. and STEWART N. G. (1960) Use of long-lived natural radioactivity as an atmospheric tracer. *Nature* **186**, 584–589.

CHERRY R. D. (1964) Alpha-radioactivity of plankton. *Nature* **203**, 139–143.

FLYNN W. W. (1968) The determination of low levels of polonium-210 in environmental materials. *Anal. Chim. Acta* **43**, 221–227.

FOLSOM T. R. (1966) Studies of background radioactivity levels in the marine environment near southern California. IMR-TR-922-66-A.

GOLDBERG E. D. (1963) Geochronology with lead-210. In *Radioactive Dating*, p. 121. International Atomic Energy Agency, Vienna.

HILL C. R. (1965) Polonium-210 in man. *Nature* **208**, 423–428.

HOLTZMAN R. B. (1963) Measurement of the natural contents of RaD (Pb^{210}) and RaF (Po^{210}) in human bone-estimates of whole-body burdens. *Health Phys.* **9**, 385.

HOLTZMAN R. B. (1966) Natural levels of lead-210, polonium-210 and radium-226 in humans and biota of the Arctic. *Nature* **210**, 1094–1097.

HOLTZMAN R. B. (1969) Concentrations of the naturally-occurring radionuclides Pb-210, Po-210 and Ra-226 in aquatic fauna. Argonne National Laboratory, Radiological Physics Division Annual Report, July 1966 through June 1967, pp. 82–89 and Proceedings of the Second National Symposium on Radioecology held at Ann Arbor Michigan, May 1967. CONF-670503.

LAMBERT G., ARDOUIN B., NEZAMI M. and POLIAN G. (1966) Possibilities of using lead-210 as an atmospheric tracer. *Tellus* **18**, 421–426.

LAMBERT G. and NEZAMI M. (1965) Determination of the mean residence time in the troposphere by measurement of the ratio between the concentrations of lead-210 and polonium-210. *Nature* **206**, 1343–1344.

LITTLE J. B. and McGANDY R. B. (1966) Measurement of polonium-210 in human blood. *Nature* **211**, 842–843.

LOEWE F. (1957) Precipitation and evaporation in the Antarctic. In *Meteorology of the Antarctic*, (editor M. P. Van Rooy). Weather Bureau, Department of Transport, Pretoria, South Africa.

PATERSON R. L., JR. and LOCKHART L. B., JR. (1964) Geographic distribution of lead-210 (Ra D) in the ground-level air. In *The Natural Radiation Environment*, pp. 383–392. Chicago University Press.

PIERSON D. H., CAMBRAY R. S. and SPICER G. S. (1966) Lead-210 and polonium-210 in the atmosphere. *Tellus* **18**, 427.

REPORT, BATTELLE-NORTHWEST (1968) Radiological Chemistry. Washington, Pacific Northwest Lab. BNWL-715-(Pt-2) (1968).

RAMA, KOIDE M. and GOLDBERG E. D. (1961) Lead-210 in natural waters. *Science* **134**, 98–99.

SHANNON L. V. (1969) The alpha-activity of marine plankton. Investl Rep. Div. Sea Fish. S. Afr. **68**, pp. 1–38.

SHANNON L. V. (1970) Oceanic circulation off South Africa. *Fish. Bull.* **6**, in press.

SHANNON L. V. and CHERRY R. D. (1967) Polonium-210 in marine plankton. *Nature* **216**, 352–353.

SVERDRUP H. U., JOHNSON M. W. and FLEMING R. H. (1942) *The Oceans, their Physics, Chemistry and General Biology*. Prentice-Hall.

SZABO B. J. (1967) Radium content in plankton and sea water in the Bahamas. *Geochim. Cosmochim. Acta* **31**, 1321–1331.

TER HAAR G. L., HOLTZMAN R. B. and LUCAS H. F., JR., (1967) Lead and lead-210 in rainwater. *Nature* **216**, 353–354.

TURNER R. C., RADLEY J. M. and MAYNEORD W. V. (1958) The alpha-ray activity of human tissues. *Br. J. Radiol.* **31**, 397–406.

VINOGRADOV M. E. (1961) Feeding of the deep sea plankton. *I.C.E.S. Symposium on Zooplankton Production*.

Food Value of Red Tide
(Gonyaulax polyedra)

Stuart Patton

P. T. Chandler

E. B. Kalan

A. R. Loeblich III,

G. Fuller

A. A. Benson

Plankton is one of the potential sources of food for expanding world needs. We studied this possibility after observing the so-called red tides which commonly occur during early summer in the coastal waters of southern California. Although economic harvesting of such minute organisms poses difficult technological problems, the hundreds of square miles of ocean area often covered by plankton blooms suggest the substantial quantities of material involved. In addition to problems of harvesting, the dominant organism in southern California red tide, *Gonyaulax polyedra*, is reputed to be toxic (*1*).

Chapman (*2*) states that anchovy and larger fish represent a satisfactory size for harvesting from the ocean but that lesser fish and plankton are too small to be recovered economically. Hence, the principle at present is to catch fish after the fish have harvested the plankton. However, because there is always a loss of efficiency during passage of nutrients along a food chain, it is important to determine the nutritive value of plankton to nonaquatic species of animals. We now describe tests of this value (particularly protein quality) in red-tide samples collected at La Jolla, California, in July 1965 and December 1966. We have reported on fatty acid composition of lipids from red tide (*3*).

Gonyaulax polyedra (*4*) was the predominant organism in our two harvests of red tide. Cells were recovered from the ocean by filtration, packed by sedimentation, and held in deep-frozen blocks for later drying to a free-flowing powder. This was done either by lyophilization or by drying of the melted slurry of cells under vacuum with a rotating evaporator.

To gain some impression of the nutritional qualities of the material, we made a proximate analysis of the sample taken in July 1965. As dry matter, the results (in percentage) were: protein, 27.5; ether extract, 4.7; fiber, 19.5; ash, 17.6; and nitrogen-free extract, 30.6. The high ash content could also be derived by assuming that the 82-percent moisture in the wet-packed cells has about the 3-percent salinity of seawater; this salinity is increased roughly fivefold by drying of the cells. Although not submitted to the complete analysis, the sample taken in December 1966 exhibited a comparable quantity of protein (26 percent as dry matter) and an amino acid composition very similar to the earlier sample.

Amino acid analyses were made in triplicate for the two plankton samples and for a culture of *G. polyedra* grown (5) in the laboratory. Each sample (1 to 3 mg) was hydrolyzed in a vacuum for 24 hours at 110°C in 6N HCl. The digested samples were analyzed by the Piez-Morris system with triangulation of the peak areas (6). The results (Table 1) are averages of the triplicates. For comparison, data on casein (7) are included. Tryptophan was determined in duplicate for two of the samples (8).

The data on amino acid composition indicate that the protein of *G. polyedra* is of good nutritional quality. This composition is quite similar to that of casein, a protein well known for high nutritional value. The three samples of *G. polyedra* (from the ocean in summer or in winter or grown in the laboratory) have very similar patterns of amino acid composition. We attribute this to strong genetic control of protein composition.

Red tide was fed to rats in two trials. The first was a simple evaluation of toxicity; when the plankton-fed rat grew about as well as its controls, we made a second trial to study nutritive quality of the material. Unfortunately, the available amount of processed red tide was of the order of only a few hundred grams; trials, therefore, were largely exploratory.

In the first trial, dried red tide (July 1965) was substituted in a purified diet for casein and cornstarch, so that the protein from red tide replaced approximately 20 to 25 percent of the protein supplied by casein in the purified diet (the control diet of the second trial with minor quantitative variations). The nonprotein fraction of red tide replaced

Table 1. Percentage of amino acids in whole casein and in three cultures of *Gonyaulax polyedra*. (A) Sample from the ocean in July 1965; (B) sample from the ocean in December 1966; (C) sample from cultures in the laboratory.

Amino acid	Whole casein	Cultures of *G. polyedra*		
		A	B	C
Asp	7.1	10.0	10.0	9.9
Thr*	4.9	4.8	4.8	4.9
Ser	6.3	4.7	4.9	4.8
Glu	22.4	12.4	12.7	13.7
Pro	10.6	5.2	5.2	5.2
Gly	2.0	5.7	6.1	7.3
Ala	3.2	7.4	7.9	8.0
½ Cys	0.3	1.0	0.7	0.9
Val*	7.2	6.5	6.4	6.4
Met*	2.8	2.9	2.9	2.7
Ileu*	6.1	4.7	4.5	4.2
Leu*	9.2	9.5	9.0	8.9
Tyr	6.3	3.9	3.6	3.3
Phe*	5.0	6.0	5.5	5.1
Lys*	8.2	6.7	7.5	6.4
His*	3.1	2.5	2.4	2.2
Try*	1.7		1.5	1.5
Arg	4.1	6.1	6.0	6.0

*Essential amino acids for the rat.

cornstarch; this caused the experimental ration of red tide to be lower in energy than the control because of the lower energy value of red tide with respect to cornstarch. Because of the small amount of red tide, we were able to test this material with only one animal. One female rat (74 g) was fed this diet for 10 days. During this period, the animal gained 3.0 g per day. This rate of gain was slightly lower than that (3.5 g per day) exhibited by comparable females on the casein control diet. The animal consumed the red tide in an amount similar to that of the control diet.

65

In the second trial, control and red-tide diets were formulated (*9*) to be equal to both protein and energy. Protein and energy values were obtained for all ingredients except minerals and vitamins. Red tide (an equal mixture of samples from July 1965 and December 1966) was added until it supplied 20 percent of the protein; the remainder was supplied by casein. Cornstarch and corn oil were adjusted so that the energy value was equal to that of the control diet.

Four male littermate rats, about 4 weeks of age, were paired and assigned to the two diets. The two animals receiving red tide grew at the same rate as those receiving the control diet. The feed intakes were about the same for both diets, and the gain for each animal during the 30-day test period was approximately 160 g. All the animals were placed on a diet of commercial rat chow for 7 days beginning on day 23. Evidently, the animals on the diet of red tide had deposited body protein rather than increased weight through large intakes of water. At the end of the trial, when the rats were killed in order to determine body fat and abnormalities, all animals seemed to have been normal, and no difference in the amount of internal fat was apparent.

Two trials for nitrogen balance were conducted during days 5 to 10 and days 16 to 23. Urine and feces, collected separately for each trial, were blended, sampled, and analyzed. The animals were in positive nitrogen balance at all times; the amount of nitrogen retained was higher during period I (60 percent) than during period II (50 percent). The digestion coefficient for nitrogen was 90 percent in period I, but it was slightly lower in period II. The animals on the diet containing plankton, as well as the control animals, seemed to utilize nitrogen during both periods. Palatability of the red tide was not a problem at any time despite marine odor and high salinity.

Our analyses of the nutrient composition of *G. polyedra* indicate its potential as a source of food. Exploratory feeding trials showed that the rats had good early growth and no toxicity during that period. Factors determining the occurrence, size, and variety of organisms of plankton blooms in the open ocean are not well known, and *G. polyedra* is not one of the more common varieties. However, controlled farming of plankton could become an integral part of the recovery (by atomic energy) of potable water and chemicals from seawater. In many arid coastal regions of the world, drying of plankton may be practical because of climates of high heat and low humidity.

References and Notes

1. J. Schradie and C. A. Bliss, *Lloydia* 25, 214 (1962); E. J. Schantz, J. M. Lynch, G. Vayvada, T. Matsumato, H. Rapoport, *Biochemistry* 5, 1191 (1966).
2. W. M. Chapman, *Food Technol.* 20, 45 (1966).
3. S. Patton, G. Fuller, A. R. Loeblich III, A. A. Benson, *Biochim. Biophys. Acta* 116, 577 (1966).
4. *G. polyedra* is a unicellular, photosynthetic, marine dinoflagellate averaging about 40 μ in diameter [see F. T. Haxo, in *Comparative Biochemistry of Photoreactive Systems*, M. B. Allen, Ed. (Academic Press, New York, 1960), p. 345].
5. Minor modifications of the culturing method of Schradie and Bliss (*1*) were used.
6. K. A. Piez and L. Morris, *Anal. Biochem.* 1, 187 (1960).

7. W. G. Gordon and E. O. Whittier, in *Fundamentals of Dairy Chemistry*, B. H. Webb and A. H. Johnson, Eds. (Avi Publishing Co., Westport, Conn., 1966), p. 60.
8. J. R. Spies, *Anal. Chem.* **39**, 1412 (1967). We thank him for the tryptophan analyses.
9. Composition (in percentage) of the control and red-tide diets, respectively, for each ingredient was: casein, 20 and 16; red tide, 0 and 17.89; cornstarch, 64 and 47.61; corn oil, 10 and 12.5; minerals, 4 and 4; and vitamins, 2 and 2.
10. We are grateful to A. E. Branding for technical assistance. Supported in part by PHS grant HE 03632. Paper No. 3271 in the Journal Series of the Pennsylvania Agricultural Experiment Station.

Red and Far-Red Light Effects on a Short-Term Behavioral Response of a Dinoflagellate

RICHARD FORWARD
DEMOREST DAVENPORT

Phytochrome mediates several long-term growth responses (flowering, seed germination, stem elongation) in higher plants (1). Studies of leaf movement in *Mimosa* (2) and in *Albizzia* (3), and of algal chloroplast movement (4), suggest the participation of this pigment in short-term light responses, not mediated by way of effects on RNA metabolism. Phytochrome, however, has not been implicated in the control of movement in motile lower plants, perhaps because the effects of prior irradiation on the light-induced behavioral response have been ignored. We now report the

possible involvement of phytochrome in a short-term response of the dinoflagellate *Gyrodinium dorsum* Kofoid.

Our culture procedure and experimental apparatus remain as reported (5) with the following modifications. Experiments were only conducted 8 hours after the beginning of a light period, on 5-day-old cultures grown on a 16-hour-light, 8-hour-dark cycle. Stimulation sources were a Bausch and Lomb No. 33-86-02 grating monochromator with a 20-nm bandpass between 400 and 740 nm, and a Toyoda microscope lamp in combination with

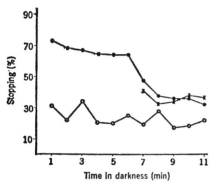

Fig. 1. Time course of typical stop response. The cells are removed from the culture lights (one 60-watt incandescent lamp plus seven daylight fluorescents) and placed in darkness at time zero. Solid circles, percentage stopping upon repeated blue stimulation (2 seconds at 470 nm from the monochromator); ×'s, response to repeated stimulation after 7 minutes in darkness; open circles, random stopping without light stimulation.

a 470-nm interference filter (Optics Technology, Inc.) having a 20-nm bandpass. Intensity from the monochromator was controlled by a continuously graded neutral-density filter (Optical Coating Laboratory). Both stimulation sources were calibrated for wavelength energy as described (5).

The response studied was the cessation of movement (stop response) upon stimulation with light of the proper wavelength and intensity. The peak in the action spectrum for this response and for oriented phototactic movement is 470 nm (5).

The stop response was recorded photographically. Since its latency was 0.4 to 0.6 second (5), the picture was taken mechanically by a solenoid-driven Robot Recorder-24 camera, 0.75 second after beginning of stimulation, to record the maximum numbers of responding cells. Stopped cells appeared round, and those moving appeared as blurry lines in ¼-second exposures. The sample size was 150 to 200 cells.

The response becomes meaningful as a behavioral assay if, under varying experimental conditions, one considers change in the percentage of cells responding plotted against time. A level of stopping above 50 percent is a positive stop response, and a level of stopping from 35 to 50 percent (as typical of response from 7th minute on) is a response drop-off (Fig. 1).

Response to 2 seconds of 470-nm light was inactivated by a longer exposure at this wavelength prior to stimulation. Cells were removed from the culture box, irradiated for 4 minutes at 470 nm from the monochromator and microscope lamp, and then placed in darkness; responsiveness was then tested as before. The response for the length of the experiment (10 minutes) was not different from random stopping. The 4-minute blue irradiation was chosen because a series of increasing exposure times showed that it always completely abolished responsiveness.

A positive stop response could be restored if the blue irradiation was followed by red irradiation. To test for the effective wavelength of reactivation, the cells were first given 4 minutes of blue, and then a 45-second exposure to light of equal energy at various wavelengths in the red. Thereafter, the cells were left in darkness and stimulated every minute at 470 nm. Exposure for 45 seconds was selected because we found that at least 30 seconds was required for maximum reactivation. The most effective wavelength for reactivation is 620 nm since it allows the positive stop response to persist for the longest time (Table 1).

Reversibility with far-red radiant energy was tested by a 45-second irradiation at wavelengths between 660 and 740 nm after 4 minutes of 470 nm and reactivation with 45 seconds at 620 nm. Cells were then left in darkness and stimulated every minute with 470 nm light. Shortening the time in darkness

69

until blue light fails to produce a positive stop response implies reversibility. The activating effects of 620 nm were most effectively reversed by exposure to 700 nm (Table 1).

In another experiment, cells were given 4 minutes at 470 nm, and then alternate 45-second exposures of red and far-red. Repeated reversibility between responsiveness and nonresponsiveness occurs, and the duration of the

Table 2. Reversibility of stop response activation and inactivation by red (620 nm) and far-red (700 nm) irradiation.

Combinations of red (R), far-red (FR) after blue exposure	Time in darkness until response drop-off (min)
FR	0
R	7
R, FR	2
FR, R	7
FR, R, FR	2
R, FR, R,	7
R, FR, R, FR	2
FR, R, FR, R	7

Table 1. Effectiveness of red and far-red light at reactivation and reinactivation of stop response, respectively. (A) Time in darkness, afer red exposures, until response drop-off. (B) Time in darkness after far-red exposures (same energy level as 620 nm) until response drop-off.

Wavelength (nm)	A (min)	B (min)
580	0	
600	4	
610	6	
620	7	
630	6	
640	4	
660	0	6
680	0	3
700	0	2
720	0	3
740	0	6

pigment rather than a chlorophyll; and the shorter wavelengths may not be unexpected in that other lower plant phytochromes show a similar shift (6).

Since red and far-red light do not produce any overt behavioral response, and the shorter wavelength absorption peaks of phytochrome in higher plants are in the long ultraviolet (7), phytochrome may be acting in combination with a blue-absorbing pigment to control the observed response. The alternative remains that there may be a phytochrome which has a blue absorption peak at 470 nm.

response is solely governed by the last red or far-red irradiation (Table 2). This reversibility by red and far-red light indicates that phytochrome is a pigment in this photoresponse of *G. dorsum.*

Under conditions of our experiments, prior irradiation with red or far-red light affects the stop response of *G. dorsum.* The effective wavelengths (620 and 700 nm) are shorter than those reported for phytochrome-mediated responses in higher plants (660 and 730 nm), and resemble the absorption regions of chlorophylls c and a. Reversibility, however, implicates a phytochrome as the active

References and Notes

1. W. S. Hillman, *Annu. Rev. Plant Physiol.* 18, 301 (1967).
2. J. C. Fondeville, H. A. Borthwick, S. B. Hendricks, *Planta* 69, 357 (1966).
3. W. S. Hillman and W. L. Koukkari, *Plant Physiol.* 42, 1413 (1967); M. J. Jaffe and A. W. Galston, *Planta* 77, 135 (1967).
4. W. Haupt, *Planta* 53, 484 (1959).
5. W. G. Hand, R. Forward, D. Davenport, *Biol. Bull.* 133, 150 (1967).
6. A. O. Taylor and B. A. Bonner, *Plant Physiol.* 42, 762 (1967).
7. W. L. Butler, S. B. Hendricks, H. W. Siegelmann, *Photochem. Photobiol.* 3, 521 (1964).
8. Supported by ONR contract Nonr-4222(03) and NSF grant GB-5137. We thank Drs. L. Blinks and B. Sweeney for reading the manuscript.

Distribution of *Gonyaulax tamarensis* Lebour in the western North Sea in April, May and June 1968

G. A. Robinson

THE deaths of sea birds, sand eels and invertebrates, and the occurrence of shellfish poisoning during May 1968 suggested there may have been a "red tide" of toxic dinoflagellates off the north-east coast of England. Red tides are caused by an abnormally high rate of reproduction of dinoflagellates, usually *Gonyaulax* spp., and are relatively rare events in British coastal waters but occur fairly frequently in tropical and sub-tropical waters off the coasts of Peru, Japan, Africa and both eastern and western coasts of the United States and Canada.

Continuous Plankton Recorders[1,2] are towed by merchant ships at intervals of a month or less along a number of standard routes at a depth of 10 m. The plankton is caught on a band of silk with sixty meshes to the inch, and preserved in a storage chamber containing formaldehyde. The silk is cut into sections, each corresponding to 10 miles of tow through the sea. The method of analysis of recorder samples has been described by Colebrook[3]. Although the recorder was designed primarily for sampling zooplankton, experience has shown that the samples reflect both the geographical distributions and fluctuations in abundance of the small organisms of the phytoplankton[4].

Plankton recorders were towed on four standard routes in the region of the toxicity (Leith–Bremen, Leith–Copenhagen, Leith–Rotterdam and Hull–Copenhagen). Large numbers of a dinoflagellate, now identified as *Gonyaulax tamarensis* Lebour, were found in the samples near the coast of north-eastern England and south-eastern Scotland, and are probably the origin of the mussel poisoning.

Fig. 1 shows the distribution of *G. tamarensis* in April, May and June 1968. The centre of each 10 mile sample has been indicated by a dot and the number of cells of *G. tamarensis* per sample ($\times 10^{-4}$) is given above each dot.

Figure 1.

Distribution of *Gonyaulax tamarensis* Lebour in April, May, and June 1968 in the western North Sea from the Firth of Forth to the Humber from samples taken by the Continuous Plankton Recorder. A key to the numbers and letters used in the chart is given below.

    • Center of 10-mile sample
1-24 Numbers per sample $\times 10^{-4}$
   X Farne Islands
   F Firth of Forth
   N Newcastle
   H Hartlepool

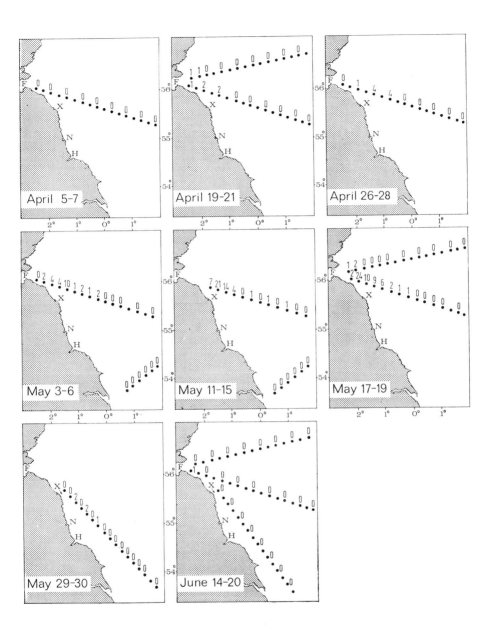

Usually every alternate sample is analysed (as in the charts for April and June in Fig. 1), but all samples were examined within the region of the outbreak during May. The cell contents of *G. tamarensis* contract into a cyst-like body when preserved; the epitheca also tends to break away from the hypotheca in the region of the girdle so that in most specimens only one part of the theca was attached to the cell and in some the theca was absent altogether. Prakash[5] also found that the cells of this species shed their thecae when they come into contact with formaldehyde or are exposed to abrupt temperature changes.

G. tamarensis first appeared east of the Firth of Forth in samples taken between April 19 and 21. Numbers increased and spread south-eastwards at the beginning of May, reaching a maximum of 240,000 cells per sample during the middle of May. Rough estimates, based on sampling tests with diatoms, suggest that this maximum represented at least 10,000 cells of *G. tamarensis*/l. Cell densities as high as $13 \cdot 54 \times 10^6$ cells/l. were found by Prakash and Taylor[6] in the Strait of Georgia. Samples taken on the Leith–Rotterdam line at the end of May showed a reduction in numbers, and had almost disappeared by the middle of June. It is possible that the recorders were not towed through the areas of maximum abundance of *G. tamarensis*, but Fig. 1 shows that it was both widely distributed and numerous in waters near the coasts where toxicity occurred.

Red tides in other areas have been caused by *Exuviella* sp., *Prorocentrum micans* Ehrenberg, *Gymnodimium breve* Davis, *Gonyaulax polyedra* Stein, *G. acatenella* Wheden and Kofoid, and *G. catenella* Kofoid, but the species of the genus *Gonyaulax* are the chief ones associated with the production of the paralytic toxin found in shellfish. In the Bay of Fundy, poisoning of shellfish was attributed to *G. tamarensis* but there was no discoloured water[5]; specimens taken in the plankton recorder survey were feebly coloured. The conditions which favoured the development of *G. tamarensis* along the east coast are not known but in previous reports (for example, refs. 7–9) red tides in other areas have been associated with higher than average water temperature, a high nutrient content, calm conditions, proximity to land and usually a lowered salinity.

I thank P. C. Wood of the MAFF Fisheries Laboratory, Burnham-on-Crouch, for a culture of *G. tamarensis* and Dr J. D. Dodge of Birkbeck College, London, for assistance in identification.

[1] Hardy, A. C., *Hull Bull. Mar. Ecol.*, **1**, 1 (1939).
[2] Glover, R. S., *Symp. Zool. Soc.* **19**, 189 (1967).
[3] Colebrook, J. M., *Bull. Mar. Ecol.*, **5**, 51 (1960).
[4] Robinson, G. A., *Bull. Mar. Ecol.*, **6**, 104 (1965).
[5] Prakash, A., *J. Res. Bd. Canad.*, **20**, 983 (1963).
[6] Prakash, A., and Taylor, F. J. R., *J. Res. Bd. Canad.*, **23**, 1265 (1966).
[7] Ryther, J. H., *Publ. Amer. Assoc. Adv. Sci. Washington*, **387** (1955).
[8] Steven, D. M., *J. Mar. Res.*, **24**, 113 (1966).
[9] Ketchum, B. H., and Keen, J., *J. Mar. Res.*, **7**, 17 (1948).

THE EFFECT OF INVASION RATE, SPECIES POOL, AND SIZE OF AREA ON THE STRUCTURE OF THE DIATOM COMMUNITY

By Ruth Patrick

Many papers have been written concerning the diversity of communities of naturally occurring species. Various diversity indices (Fisher, 1943; Shannon and Weaver, 1948) have been formulated, and various types of models (Preston, 1948; MacArthur, 1957) have been used to describe the structure of the community in terms of its diversity. More recently, MacArthur and Wilson (1963) have emphasized the importance of size of area and invasion rate in the maintenance of a diversified community.

In the present studies three series of experiments were performed. One was to show what effect the size of area and the number of species in the species pool which were capable of invading an isolated area had on the numbers of species which composed the diatom community. The second series of experiments was to show what effect altering the invasion rate had on the structure of the community. The third series of experiments was a comparison of the structure of diatom communities in structurally and chemically similar streams on the island of Dominica and in the United States. The temperature of the water in the two rivers was similar when the studies were made.

Methods and Procedures.—In order to carry out the first series of experiments, duplicate small glass squares 9 mm², 36 mm², and 625 mm² were each erected on a small plastic pedicel about 10 mm in length which was fastened onto a glass slide. These glass slides were placed in plastic boxes which were opened so that the current would pass across the glass squares, giving similar conditions for the attachment of the diatoms across the surface. Having the current strike the slides would have created strong, uneven current patterns.

The slides were placed in clean boxes each day and the pedicels and slides carefully cleaned. This was to minimize the chance that diatoms from sources other than the passing water might invade the glass squares on the ends of the pedicel. The experimental design was to simulate the invasion of species from distant sources onto islands of varying size ranges. These slides were placed in the flowing water from a spring (Roxborough Spring) and in a eutrophic stream (Ridley Creek). Previous studies had shown that the total number of species in Roxborough Spring in the area studied was about 60 at any one time, and in Ridley Creek about 250. It should be noted that the chemical characteristics of the water and the temperature were more variable in Ridley Creek than in Roxborough Spring. Since the spring has a deep source, the chemical characteristics and temperature of the water in it remain almost constant over long periods of time.

The second series of experiments was designed to simulate what might happen if the invasion rate were lowered as happens on an island fairly distant from a highly diversified species pool, such as pools that exist on continents. An island of any significant size would have almost infinite size as far as the diatom community was concerned. The main factor limiting diversity would be the invasion rate.

In this series of experiments, as contrasted with the first series, the invading species in each case were from the same water and the same species pool was available, but the rate of invasion was different.

Glass slides (3 in. $\times$ 1 in.) were placed in boxes and water flowed over their surfaces at 550–600 liters/hour. This rate of flow had previously been found to produce on such slides diatom communities which were typical of a natural stream. In another box the same rate of flow was maintained by recycling filtered stream water which furnished most of the flow. During the first three days diatoms were allowed to invade at a rate of 550–600 liters/hour. The reason for allowing the full flow during the first few days was to develop some semblance of a diatom flora on the slides. It is usually about two days before the diatom flora starts to develop. After the first three days, the invasion rate of new diatoms was at 1.15 liters/hour. Two sets of these experiments were carried out in a eutrophic stream, one in late September to early October, and one in late October to early November.

The third series of experiments was carried out under natural conditions. Diatom communities were studied in similar oligotrophic streams on the island of Dominica and in the state of Maryland.

Discussion of Results.—The results of the first series of experiments are shown in Table 1*A*. These experiments show that size of area influences the number of species established in an area, that the number of species increases greatly at first and is not very different between four days and one week, and that subsequently the pattern of increase is irregular. On one of the 36-mm² slides exposed in the fall in Roxborough Spring there seems to be a slight decrease of species at eight weeks. Other experiments carried out in a similar manner in Roxborough Spring in the summer, when growth is many times as rapid, showed this decrease more distinctly at the end of two weeks (Table 1*B*). In each case the species which disappeared were those represented by very small populations.

TABLE 1

(*A*) Experiments in September-October 1964

Size of slide:		Roxborough Spring					Ridley Creek 36 mm² No. of species, box 7
	625 mm²		36 mm²		9 mm²		
			Number of Species				
	Box 1	Box 2	Box 3	Box 4	Box 5	Box 6	
4 days	46	37	23	23	1	3	—
1 week	40	32	28	24	7	—	—
2 weeks	54	35	—	22	10	10	—
8 weeks	—	—	29	14	19	14	160

(*B*) Experiments in Roxborough Spring during Summer 1964

	1 Week, 144-mm² slide		2 Weeks, 144-mm² slide		1 Week, 625-mm² slide		2 Weeks, 625-mm² slide	
	Box 1	Box 2	Box 3	Box 4	Box 5	Box 6	Box 7	Box 8
No. of species	32	28	23	22	47	44	29	28

When we compare slides of the same size (36 mm²) from Roxborough Spring and from an area in Ridley Creek which had a much larger species pool and a more rapid invasion rate because of swifter current, we see that at the end of eight weeks 160 species were established on the Ridley Creek slides whereas 14–29 species were established on the Roxborough Spring slides (Table 1*A*).

The second series of experiments in which the invasion rate was reduced but the area was large produced the results shown in Figures 1–4. The number of species composing the community was reduced and the sizes of the populations were more variable than in the communities with a high invasion rate. This is shown by the reduction in the height of the mode, the reduction in the observed and calculated species in the community, the increase in σ^2, and the number of intervals covered by the curve. The diversity index (Shannon and Weaver, 1948) is also less. This decrease in the diversity index, particularly in the October–November study, is due to the fact that two species, *Nitzschia palea* and *Navicula luzonensis*, are excessively common in the reduced-flow-rate community. In the September–October reduced-flow study, the commonest species is represented by about one seventh the number of specimens as in the October–November reduced-flow experiments, and more species are represented by fairly large populations. Therefore, the diversity index is not as greatly reduced. This difference in sizes of populations of species is probably due to the fact that conditions for growth were better for more species in the September–October period than in the October–November period.

When one compares the total biomass of the September–October community that had a lower invasion rate and lower number of species with the biomass of the community that had a higher invasion rate and larger number of species, it is only 17 per cent greater in the community with fewer species (21.2 mg as compared to 18 mg). This is probably within the range of natural variation, and the biomasses of the two communities are not really different. It would appear that the numbers of species composing the communities does not significantly affect the total biomass of the community.

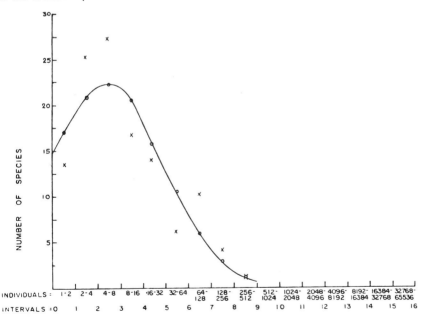

Fig. 1.—Invasion rate 550–600 liters/hr, October–November, 1964. Height of mode, 22.4 species; observed species, 123; σ^2, 6.2; intervals covered by the curve, 9; diversity index, 3.805.

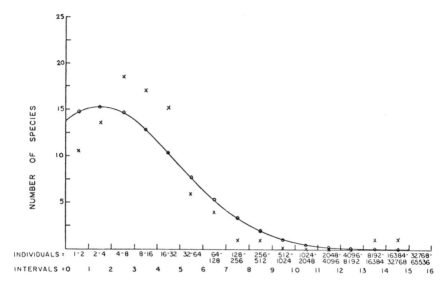

FIG. 2.—Invasion rate 1.5 liters/hr, October–November, 1964. Height of mode, 15.3 species; observed species, 97; σ^2, 12; intervals covered by the curve, 15; diversity index, 0.972.

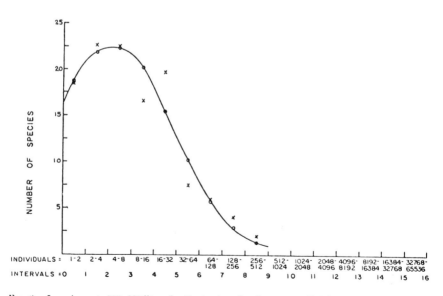

FIG. 3.—Invasion rate 550–600 liters/hr, September–October, 1964. Height of mode, 22.5 species; observed species, 129; σ^2, 6.9; intervals covered by the curve, 9; diversity index, 3.713.

78

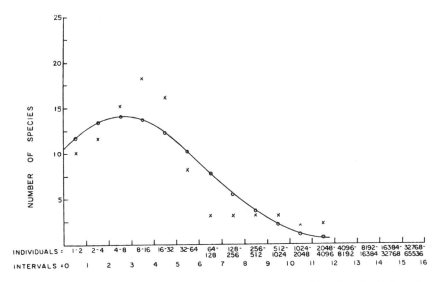

Fig. 4.—Invasion rate 1.5 liters/hr, September–October, 1964. Height of mode, 13.9 species; observed species, 100; σ^2, 12.6; intervals covered by the curve, 12; diversity index, 2.522.

The results of the third series of studies are set forth in Figures 5–7. It will be seen that the number of species in the two Dominica diatom communities is considerably smaller than in the diatom community in the United States. The sizes of the populations of the most common species in all three communities fell in the same interval of the truncated normal curve and were represented by 14,400 specimens in the Layou River, 9,525 specimens in Check Hall River, and 15,975 specimens in Hunting Creek. The main difference in the population sizes of the various species in the streams was that although there were fewer species in the Dominica streams, more of them had fairly large populations than those in Hunting Creek. Most of the species in Hunting Creek had moderate to very small populations. Therefore σ^2 is larger in the Dominica streams than in Hunting Creek. Likewise the Shannon-Weaver diversity index is larger in the Dominica streams. However, Fisher's α (1943), which effectively indicates species numbers even if the populations are small, is much larger in Hunting Creek (25.99) than in the Dominica streams (Layou River, 5.594; Check Hall River, 5.667). This latter index seems more clearly to indicate the differences in species numbers in these two types of communities, while the Shannon-Weaver index more clearly indicates the unevenness of the distribution of individuals in the various species populations.

Conclusions.—From these studies, which were controlled, semilaboratory experiments as well as actual field studies, it is evident that size of area, number of species in the species pool which are capable of invading the area, and the rate of invasion by the species greatly influence the numbers of species and the diversity of the community.

A reduced invasion rate (size of area and number of species in the species pool remaining the same) reduced the total number of species in the community, par-

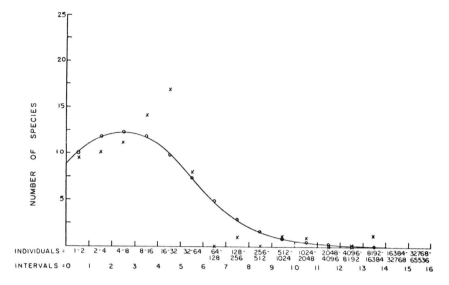

Fig. 5.—Hunting Creek, Maryland; truncated curve for a diatom community. Height of mode, 12 species; observed species, 79; σ^2, 9.1; intervals covered by the curve, 14; diversity index, 0.789.

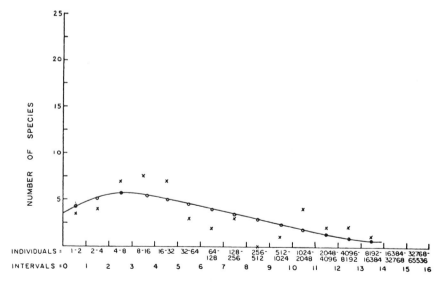

Fig. 6.—Check Hall River, island of Dominica; truncated curve for a diatom community. Height of mode, 5 species; observed species, 46; σ^2, 21.6; intervals covered by the curve, 14; diversity index, 1.919.

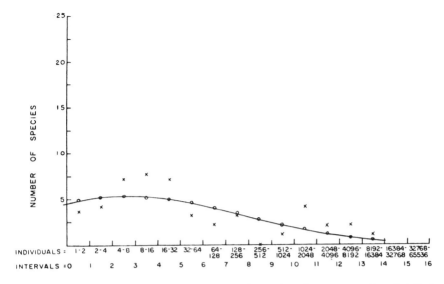

FIG. 7.—Layou River, island of Dominica: truncated curve for a diatom community. Height of the mode, 5 species; observed species, 49; σ^2, 26; intervals covered by the curve, 14; diversity index, 2.028.

ticularly those species with small populations which are typically part of a natural continental community. There is an increase in species with fairly large populations although the total number of species decreases.

Similar results were found when diatom communities on the island of Dominica were compared with a diatom community in a similar type of stream in the state of Maryland. The island communities had much smaller numbers of species with larger populations than the continental community. They also had fewer species with small populations.

One of the main results of a high invasion rate is to maintain in a community a number of species with relatively small populations. The value of these species to the community may be similar to the presence of relatively rare genes in a gene pool. They may function to preserve the diversity of the community under variable environmental conditions. Such rare species at any one point of time under changed environmental conditions might be better adapted than the presently common species and might be able to increase rapidly.

The size of the species pool which may potentially invade an area also has a great effect on the number of species composing the community, as seen in the comparison of the species numbers in the communities in the Roxborough Spring studies and the Ridley Creek studies. The size of the area to be invaded also affects the diversity of the community. Once the area becomes filled, the first species to be eliminated were those represented by very small populations.

The author wishes to express her gratitude to Dr. G. E. Hutchinson and Dr. Robert MacArthur for their helpful advice and criticism, and to Miss Noma Ann Roberts, Mr. Raymond Cummins, Mr. Roger Daum, and Miss Lee Townsend, who have been most helpful in the carrying out of these studies. The author also wishes to thank the U.S. Public Health Service (WP-00475) for their financial support for part of this work.

Fisher, R. A., A. S. Corbet, and C. B. Williams, "The relation between the number of species and the number of individuals in a random sample of an animal population," *J. Animal Ecol.*, **12**, 42–59 (1943).

MacArthur, R., "On the relative abundance of bird species," these PROCEEDINGS, **43**, 293–295 (1957).

MacArthur, R., and E. O. Wilson, "An equilibrium theory of insular zoogeography," *Evolution*, **17**, 373–387 (1963).

Preston, F. W., "The commonness and rarity of species," *Ecology*, **29**, 254–283 (1948).

Shannon, C. E., and W. Weaver, *The Mathematical Theory of Communication* (Urband: University of Illinois Press, 1948), pp. 3–91.

82

Freshwater Primary Production by a Blue–Green Alga of Bacterial Size

Loch Leven, Kinross, Scotland (National Grid reference No. 145020; latitude 56° 15′ N, longitude 3° 25′ W), is a shallow productive lake 13 km² and of 5 m mean depth. An investigation of its ecology and production of phytoplankton is being carried out as part of a chemical, physical and biological study of the lake as part of the International Biological Programme. A minute rod-shaped organism occurred as an important part of the phytoplankton between March and August 1968, and at times was present in an almost pure stand, giving a green colour to the water. A morphological investigation has been made to determine whether it is a bacterium or a blue–green alga, and its temporal abundance and spatial distribution have been studied, together with measurements of gross photosynthetic production.

Cells of natural and cultured material are pale green, almost colourless straight rods 6–40 μm long and 0·7–0·9 μm in diameter, shrinking to about 0·6 μm across when fixed and embedded. At each end there is a slight trace of a calyptra. Commonly the cells appear singly or occasionally in pairs linked end to end, although, in cultures, chains of up to four cells have been seen. With light microscopy little internal structure is discernible either with transmitted light or with phase contrast illumination. A granular and vacuolar appearance is developed to a varying extent, apparently associated with various factors including the age of the culture or the natural population. No mucilaginous sheath or gas vacuoles are apparent.

Cells from a water sample containing an almost pure population were concentrated by membrane filtration and centrifugation, fixed in 2 per cent osmium tetroxide in phosphate buffer at pH 7·0, and embedded in 'Epon'. Sections were stained in uranyl acetate and Reynolds lead stain. The microanatomy showed conclusively that the organism is a blue–green alga of bacterial proportions (Fig. 1). In transverse sections there is usually a single peripheral photosynthetic lamella, while longitudinal sections show most of the inclusions demonstrated by Pankratz and Bowen[1] in the much larger *Symploca muscorum*. In addition, at the end of each cell there are cylindrical bundles of microtubules each about 15 nm in diameter and 300 nm long, a feature apparently unrecorded so far in blue–green algae. A full account of the fine structure will be published later, together with a formal taxonomic description, for this seems to be an undescribed species of the genus *Synechococcus*. *S. plankticus* Drews,

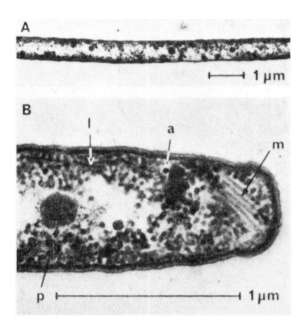

Fig. 1. Electron micrographs of *Synechococcus* cells from Loch Leven. *A*, Part of a cell (× 10,000); *B*, end of another cell (× 50,000). a, Alpha-granule; l, photosynthetic lamella; m, microtubule; p, polyhedral body.

Prauser and Uhlmann is similarly minute, but of a different shape, and has been found in quantity growing in a pool of sewage effluent[2].

The phytoplankton standing crop has been estimated regularly, usually at weekly intervals. Samples used for quantitative work were collected with a weighted tube as used by Lund[3], or with a Ruttner water bottle. Chlorophyll *a* was estimated spectrophotometrically, after filtering water samples through fine glass-fibre filters (Whatman GF/C), extracting pigments in 90 per cent methanol, and using the abbreviated equation proposed by Talling and Driver[4] to relate chlorophyll *a* concentration and optical density measured at 665 nm. Larger phytoplankton algae were counted with an inverted microscope after iodine sedimentation and smaller ones by means of a simple counting chamber[5,6] or a haema-cytometer, the two giving good agreement in counts of the *Synechococcus*.

Gross photosynthetic production has been measured at fortnightly intervals, using suspended light and dark bottles and Winkler oxygen estimations as described by Talling[7]. The euphotic zone was more or less uniform with respect to temperature and algal population density

84

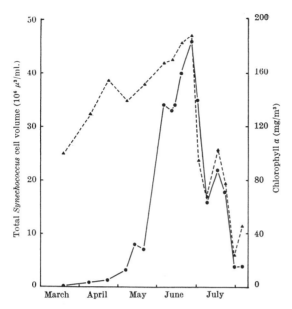

Fig. 2. Variations in the total cell volume of *Synechococcus* per ml.
(———), together with the mean chlorophyll *a* concentration (– – –),
in the upper 4 m of Loch Leven.

to a lower limit varying between 1·5 and 2·5 m depth,
and so the bottles were filled with 0·5 m samples. They
were then exposed at a series of depths for given times,
from 2 to 3 h, near midday, exposure beginning immediately after they were filled. Planimetric integration of the
depth profiles of photosynthetic productivity per unit
volume of water, in mg O_2/m^3 h, yields estimates of this
productivity per unit area of lake surface, in mg O_2/m^2 h.
Changes in the cell abundance of *Synechococcus*, concentration of chlorophyll *a* ($=n$), the maximum (light-
saturated) rate of phytosynthesis per unit of chlorophyll *a*
($=P$) and the gross photosynthetic production per unit
area ($=\Sigma nP$) are shown in Figs. 2 and 3 for the period
April to July 1968 inclusive.

Synechococcus was present in small amounts (less than
$0·2 \times 10^6$ μm³/ml.) throughout the winter and began a
gradual rise in late April, when it constituted about 3 per
cent of the total phytoplankton live volume of $36 \times$
10^6 μm³/ml., the rest consisting of diatoms. The decline
of the diatom populations accompanied the greatest rates
of increase of the *Synechococcus* in late May and early
June, and in June and July this alga accounted for more
than 90 per cent of the estimated total live crop volume.
At its maximum, on June 26, it represented about 98 per

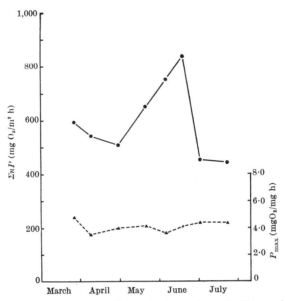

Fig. 3. The integral hourly photosynthesis per unit area (ΣnP) (———),
and maximum, light-saturated, rate of photosynthesis per unit of
chlorophyll a (P_{max}) (– – –).

cent, with 45×10^6 $\mu m^3/ml$., and then a value of 4 μg of
chlorophyll a/mm^3 (that is, 10^9 μm^3) cell volume was
found, while other estimations during June and July
gave similar values, in broad agreement with the results
of other workers[8,9]. The only large departure from this
value is that obtained on July 1, 2·7 μg of chlorophyll
a/mm^3 cell volume, and because this occurred at the
beginning of a period of population decline it may represent some true change in pigment content connected with
this decline. By the end of July the standing crop volume
of *Synechococcus* was at about one tenth of its maximum,
but it still represented 90 per cent of the total crop, the
remainder being chiefly small Chlorophyceae. From Figs.
2 and 3 it is apparent that the increase in the rate of
photosynthesis per unit area in late April and its subsequent decrease in the second half of June are more
closely correlated with changes in chlorophyll a concentration than with those in the maximum specific activity
of the chlorophyll, which remained relatively constant
throughout the period reviewed here.

There is evidence that, with increasing eutrophication
of the loch, blooms have become more common and long
lasting, and those composed of blue–green algae have
been associated with poor fishing in this famous trout

water. Whether the present species formed part of those blooms is unknown, for it could easily have been mistaken for a bacterium by the untrained observer. Occasional reports from elsewhere of plankton composed of very small blue–green algae have appeared[2], and some of the green bacteria which have been described undoubtedly belong to the former group[10-12]. It must, however, be unusual for a blue–green alga of this sort of size to occur as a pure stand in conditions where it can be studied intensively. This species is now growing in culture, and so further studies may show what influence the unusual features of size and morphology have on the rate of photosynthesis and on other aspects of metabolism and growth, as well as the production and diffusion of extra-cellular substances.

We thank various members of the Nature Conservancy and the Freshwater Biological Association for helpful comments, especially Mr N. C. Morgan, Dr J. W. G. Lund and Dr J. F. Talling. We also thank Professor I. Manton of the University of Leeds for the use of an electron microscope.

A. E. BAILEY-WATTS
M. E. BINDLOSS
J. H. BELCHER

[1] Pankratz, H. S., and Bowen, C. C., *Amer. J. Bot.*, **50**, 387 (1963).
[2] Drews, G., Prauser, H., and Uhlmann, D., *Arch. Mikrobiol.*, **39**, 101 (1961).
[3] Lund, J. W. G., *J. Ecol.*, **38**, 1 (1950).
[4] Talling, J. F., and Driver, D., *Proc. Conf. Primary Productivity Measurement, Marine and Freshwater*, Hawaii, 142 (US Atomic Energy Commission, 1963).
[5] Lund, J. W. G., *Limnol. Oceanog.*, **4**, 57 (1959).
[6] Lung, J. W. G., *Limnol. Oceanog.*, **7**, 261 (1962).
[7] Talling, J. F., *New Phytol.*, **56**, 29 (1957).
[8] Talling, J. F., *Intern. Rev. Ges. Hydrobiol. Hydrogr.*, **50**, 1 (1965).
[9] Parsons, T. R., *J. Fish. Res. Board Canad.*, **18**, 1017 (1961).
[10] Pringsheim, E. G., *Arch. Mikrobiol.*, **19**, 353 (1953).
[11] Van Niel, C. B., and Stanier, R. Y., in Ward, H. B., and Whipple, G. H., *Fresh-water Biology*, second ed. (edit. by Edmondson, W. T.), 16 (John Wiley and Sons, New York and London, 1959).
[12] Hutchinson, G. E., *A Treatise on Limnology*, **2** (John Wiley and Sons, New York and London, 1967).

Strontium-90 Concentration Factors of Lake Plankton, Macrophytes, and Substrates

Z. KALNINA

G. POLIKARPOV

Among environmental factors caused by man, the effects of man-made radionuclides are unique (*1*). Despite previous work on radioecology of aquatic organisms (*1, 2*), accumulated knowledge is insufficient for solving hydrobiological problems resulting from the use of atomic energy. One such problem involves the role of living and inert lake components in the cycling of strontium-90. Concentration factors (CF), that is, the ratio of a radionuclide in a lake component to that in water, of ^{90}Sr have been studied in the laboratory in planktonic crustacea from Lake Bolshoe Miasovo (*3*). Accumulation of ^{90}Sr by unicellular and filamentous algae (*4*), by *Cladophora* sp. and *Myriophyllum* sp. in English lakes (*5*), and by some freshwater plants (*6*) has been studied. We have now determined CF of ^{90}Sr of plankton (Tables 1 and 2), macrophytes (Table 3), and substrates (Table 4) in three different types of Latgalian lake (eutrophic, mesotrophic-eutrophic, and dystrophic). These types are represented by Rušonu, Gusenu, and Pitelu, respectively. Rušonu and Gusenu contain calcium bicarbonate, and zooplankters predominate in the plankton (Table 1). Zooplankton and phytoplankton were present in about equal amounts in Pitelu (Table 2).

Water samples were collected once a month from five typical sites in each lake. At each site two water samples (100 liters each) were taken simultaneously (*7*). Plankton was collected at the same sites with a fry-neuston trawl (sieve size No. 61) designed by Zaitsev (*8*). For radiochemical analyses, 4 to 13 grams of ashed plankton were used.

Table 1. Strontium-90 concentration factors from a eutrophic lake, 1967.

| Sample size | Mean concentration factors | | Density of zooplankton | | Dominant zooplankters |
	Ash weight	Wet weight	Count $(10^3/m^3)$	Weight (g/m^3)	
			May		
4	1,298 ± 618*	18 ± 10	470	3.4	*Bosmina coregoni gibbera, B. longirostris, Chydoris sphaericus*
			June		
9	705 ± 81	14 ± 5	326	3.7	*B. coregoni gibbera, C. sphaericus, Daphnia cucullata*
			July		
8	1,374 ± 450	20 ± 6	289	4.2	*B. crassicornis, B. coregoni gibbera, C. sphaericus*
			August		
10	385 ± 85	11 ± 1	186	3.4	*Sida crystallina, B. crassicornis, B coregoni gibbera*
			September		
10	595 ± 152	11 ± 3	240	3.0	*B. coregoni gibbera, Chydorus sp., B. crassicornis*

* Ninety-five percent confidence limits

After the addition of carriers (100 mg of Sr, 30 mg of Ce, and 30 mg of Fe), the ash was dissolved with 12N HCl and distilled water at 200°C. Undissolved residue was removed, and ammonium oxalate was added to the filtrate. The solution was then warmed and adjusted to pH 4 with ammonium hydrate and complete precipitation was verified. The oxalate precipitate was filtered and again precipitated; it was then dried and ashed at 600°C for 1.5 hours. The carbonates obtained were dissolved in 6N HCl; hydroxides were precipitated at pH 7 to 8. The filtrate was adjusted to pH 4, and the solution allowed to establish an equilibrium between ⁹⁰Sr and ⁹⁰Y. Radiochemical analyses of the ash of freshwater plants and substrate were similarly done. After equilibrium between ⁹⁰Sr and ⁹⁰Y occurred, carbonate precipitates (from water, plants, and substrates) were dissolved in 6N HNO₃ or HCl, stable yttrium was added to the solution, and the hydroxide of yttrium was precipi-

tated by ammonia (pH 7 to 8), 25 percent carbon free. The precipitate was filtered, precipitated again, and ashed; ionizing radiation was measured by a meter with a relative error of 10 percent. In the same way, ⁹⁰Y was isolated from planktonic samples after equilibrium between ⁹⁰Sr and ⁹⁰Y was established. The disintegration of isolated ⁹⁰Y was examined each time, and it always coincided with the theoretical disintegration curve for ⁹⁰Y.

Stable strontium was determined by flame photometry (9). The initial solutions were prepared by the method of addition and read in the flame of a mixture of acetylene with air at the wavelength of strontium. Ninety-five percent confidence intervals for means were calculated where sample numbers permitted.

Concentration factors of ⁹⁰Sr for plankton of eutrophic (Table 1) and mesotrophic-eutrophic (10) lakes are similar, but are greatly different from those for plankton of a dystrophic lake

89

Table 2. Strontium-90 concentration factors of plankton from different sites of a dystrophic lake, 1967 (single samples).

Concentration factor		Zooplankton		Dominant zooplankters and other components
Ash weight	Wet weight	$(10^3/m^3)$	(g/m^3)	
		May		
10,300	226	272	4.5	*Asplanchna* sp., *Bosmina coregoni gibbera*, sapropel, peat
7,933	209	136	2.4	*B. coregoni gibbera, Asplanchna* sp., sapropel
7,033	88	179	7.6	*B. coregoni gibbera*
		June		
18,571	100	67	2.8	*B. coregoni gibbera, Sida crystallina*, sapropel
12,571	303	94	6.2	*S. crystallina, B. coregoni gibbera*, peat
		July		
7,000	133	125	2.5	*S. crystallina, Chydorus sphaericus*, much phytoplankton
4,640	25	130	13.4	*S. crystallina*, much phytoplankton
7,600	118	66	2.4	*S. crystallina, Leptodora kindtii*, much phytoplankton
14,000	157	45	0.7	*S. crystallina*, much phytoplankton and sapropel
4,560	21	115	10.0	*S. crystallina*, phytoplankton
		August		
12,929	67	158	4.4	*S. crystallina, Bosmina* sp., *Asplanchna* sp., much phytoplankton and sapropel
13,161	88	93	6.7	*S. crystallina*, sapropel, phytoplankton
12,285	59	59	2.7	*S. crystallina*, very much phytoplankton, sapropel
9,910	49	221	3.1	*Asplanchna* sp., *S. crystallina*, much phytoplankton, sapropel
10,268	38	141	3.1	*Bosmina* sp., phytoplankton
		September		
8,538	21	51	1.2	*B. coregoni gibbera*, phytoplankton
13,711	84	82	4.4	*S. crystallina*, much sapropel and peat
12,807	60	91	2.3	*B. coregoni gibbera*, much sapropel and peat
13,404	63	62	4.2	*S. crystallina, B. coregoni gibbera*, sapropel
10,615	44	70	1.2	*B. coregoni gibbera*, much phytoplankton

(Table 2). This difference may be explained (*11*) in the following way: concentrations of carriers in samples from the dystrophic lake (1966—Ca, 3.6 to 6.1 mg/liter; 19~7—Ca, 4.7 to 5.9 mg/liter, and Sr, 13.6 to 43 μg/liter) are much less than those for samples of the eutrophic lake (1966 —Ca, 35.3 to 40.6 mg/liter, and Sr, 71 μg/liter; 1967—Ca, 43.8 to 46.3 mg/liter, and Sr, 67.6 to 80.6 μg/liter) and the mesotrophic-eutrophic lake (1966—Ca, 36.1 to 38.9 mg/liter).

The ^{90}Sr CF for water plants are also higher in the dystrophic lake than in the other lake types (Table 3). The ^{90}Sr CF's for water plants are less in spring than in autumn (3 to 7 times,

Table 3. Strontium-90 and stable strontium concentration factors (CF) of aquatic plants from eutrophic and mesotrophic-eutrophic lakes during September 1966 and 1967.

Plant	Mean concentration factors	
	Wet weight	Ash weight
Chara sp.	276 ± 20*	2,867 ± 604
Equisetum sp.	75 ± 33	2,053 ± 601
Scirpus lacustris	24 ± 6	1,350 ± 740
Myriophyllum sp.	63 ± 19	3,294 ± 1,185
Stratiotes aloides	96 ± 35	3,677 ± 1,005
Elodea canadensis†		3,225 ± 2,098
Ratio $CF_{max} : CF_{min}$	> 10	< 3

* Ninety-five percent confidence limits.   † Eutrophic lake only.

Table 4. Strontium-90 concentration factors for different lake substrates during September 1966.

Substrate	Concentration factors	
	Ash weight	Dry weight
Mesotrophic-eutrophic		
Silt	33–51	31–50
Sand with salt	28	26
Sand	11–14	10–14
Eutrophic		
Silt	54–9(	50–83
Silt with sand	38	29
Dystrophic		
Peat	1,097–6,386	274–1,250
Peat with sand	2,536–12,691	1 725–2,067
Sand with peat	441–528	412–493
Sand with silt and peat	956	825

calculating on the basis of wet weight). The ^{90}Sr CF are highest in the upper part of *Potamogeton lucens* (100 percent), average in the central part (50 percent), and least in the lower parts (including roots) (28 percent). These differences are noticed in autumn as well as in spring. The CF of stable Sr and ^{90}Sr are close—the mean ratio of Sr CF to ^{90}Sr CF in each sample of 9 species of plants equals to 1.1 ± 0.2. This means that complete exchange occurred between radioactive and stable Sr in lake plants.

Among lake substrates, sand substrates have the least ^{90}Sr CF (Table 4). The higher CF of sand substrates are a result of the admixture of silt and peat. The highest ^{90}Sr CF are of peat substrates from the dystrophic lake (Table 4). Concentration factors for ashed plankton in the eutrophic lake are significantly higher than those for ashed lake silt, but CF for ashed plankton and peat in the dystrophic lake are approximately equal.

Thus hydrobionts and substrates of a dystrophic lake are characterized by ^{90}Sr CF higher than those for mesotrophic-eutrophic and eutrophic lakes.

References and Notes

1. G. G. Polikarpov, *Radioecology of Aquatic Organisms* (North-Holland, Amsterdam, 1966), p. xxviii.
2. A. W. Klement, Jr., and V. Schultz, *Terrestrial and Freshwater Radioecology: A Selected Bibliography*, AEC Rep. TID 3910 (plus supplements 1–6) (U.S. Atomic Energy Commission, Washington, D.C., 1962); A. W. Klement, Jr., C. F. Lytle, V. Schultz, *Russian Radioecology: A Bibliography of Soviet Publications of English Translations and Abstracts*, AEC Rep. TID-3915 (U.S. Atomic Energy Commission, Washington, D.C., 1968).
3. E. A. Timofeeva-Resovskaya, *Trudy instituta biologii, Sverdlovsk, Akademiya Nauk SSSR, Ural'skiy Filial* **30**, 1 1963) (English transl. JPRS 21,816).
4. E. A. Gileva, in *Radioaktivnie Izotopi v Gidrobiologii i Metodi Sanitarnoi Gidrobiologii*, V. I. Zhadin, Ed. (Izd. "Nauka," Moscow-Leningrad, Academy of Science, Zoological Institute, 1964), pp. 17–20.
5. W. L. Templeton, in *The Effects of Pollution on Living Material, Symposium of the Institute of Biology No. 8*, W. B. Yapp, Ed. (Institute of Biology, London, 1959), p. 125.

6. M. Merlini, F. Girardi, G. Pozzi, in *Nuclear Activation Techniques in the Life Sciences* (Proc. of the Symp., Amsterdam, 8–12 May 1967) (International Atomic Energy Agency, Vienna, 1967), pp. 615–629.

7. G. A. Sereda and I. I. Bobovnikova, *Radiochemical Method of Mass Control of Content of Sr^{90} in Fresh Water* (Gos. Komitet Po Ispol'zovaniju Atomnoj Energii SSSR, Moscow, 1963), 12 pp.

8. Yu. P. Zaitsev, *Naukovi zapiski Odesskoy biologichnoy stantsii. Kiev, Akademiya Nuuk Ukr. SSR* **4**, 19 (1962).

9. N. S. Poluektov, *Methods of Analyses on Photometry of Flame* (in Russian) (Gos. izd. khimicheskoy literatury) (State Publishing House of Chemical Literature, Moscow, 1959), 231 pp.

10. Z. K. Kalnina and G. G. Polikarpov, *Radiobiology* **8**, 1 (1968).

11. D. C. Pickering and J. W. Lucas, *Nature* **193**, 1046 (1962).

12. We thank Prof. Yu. P. Zaitsev for consultation and help; I. A. Sokolova, A. A. Bachurin, and D. D. Ryndina for advice on determinations of ^{90}Sr and stable Sr in freshwater; D. S. Parchevskaya for assistance in statistical treatment of the data; and Prof. V. Schultz for his kind cooperation in preparation.

92

Planktonic Foraminifera: Field Experiment on Production Rate

WOLFGANG H. BERGER
ANDREW SOUTAR

Planktonic Foraminifera are of interest and value in studies of marine
zoogeography and paleoecology. Little
is known, however, about their life
cycles and productivity. The hypothesis that reproduction of some planktonic Foraminifera takes place at
great depth, expressed by Walther in
1893 (*1*), has recently been considered
proven (*2*) based on the occurrence
of heavily encrusted living foraminif-

era at depths of more than 500 m
in the Atlantic. It has been suggested
that in the North Atlantic *Globorotalia truncatulinoides* reproduces below a depth of 500 m during November, which implies a yearly cycle of
submergence and reproduction for
this species (*3*).

However, if one assumes an annual
overturn for planktonic foraminifera,
their production of empty shells was

found to be inadequate by a factor of 10, when the budget for river influx and ocean sedimentation of calcium carbonate on a worldwide basis was examined (4). It therefore seemed problematic whether annual submergence and reproduction could be extrapolated to all or even most species.

We attempted to obtain evidence bearing directly on this problem by simultaneously ascertaining both the concentration of planktonic foraminifera in the water column by net hauls and the output of this population by collecting falling tests in a sediment trap. At the same time the physical characteristics of the water column were measured from the surface to the bottom (Fig. 1). The location of the experiment was in the center of the Santa Barbara Basin off Southern California. Here the bottom is shallow, and therefore the vertical distance traveled by the shells, as well as their lateral displacement during sedimentation, is minimized. The surface circulation appears to be rotational, so that the system may be regarded as semiclosed for time spans of a few days. Thus it seems likely that the empty tests falling to the bottom represent the production in the water column above.

We assumed a steady-state condition for the concentration of living foraminifera and their empty shells in the water column during the experiment. Under these conditions the output of empty shells measured on the ocean floor equals the input from perishing individuals above. In addition, the rate of reproduction must balance the rate of attrition in the living population in order to maintain the population at the given density. Thus the turnover time may be calculated by dividing the standing crop of living foraminifera above a given area of the bottom by the number of empty tests deposited on this area during a specified time. Basically, this

is the same reasoning that Bramlette used (4).

In order to obtain valid results with this approach, the calculations would have to be made separately for each species and life stage. For a first approximation, the size of foraminifera may be taken as a measure of the life stage, so that life spans should be calculated for narrow ranges in size. We took 150 μ as a minimum adult size and treated all specimens larger than this on an equal basis, which results in turnover times that are longer than average life spans. Many foraminifera grow to a size somewhat larger than 150 μ, and their density in the water column is consequently smaller, since large specimens are rarer than small ones in the surface water. This leads to a smaller standing crop

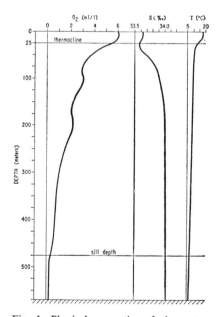

Fig. 1. Physical properties of the water column in Santa Barbara Basin during the last part of August 1966.

94

Table 1. Density profile of planktonic foraminifera in the open-closed net tows, expressed as the number of specimens per cubic meter, and the catch of empty tests in the sediment trap, expressed as the number per square meter per day. The total catch during 4 days was 73 specimens, 3 of which were smaller than 150 μ. Between 250 and 500 m^3 of water was filtered in each tow.

Species	Number per cubic meter at depths (m) of					Number per square meter per day
	0–35	35–45	45–100	100–300	300–500	
Globigerina bulloides	9.7	4.0	2.2	0.3	0.3	8.6
Globoquadrina eggeri	5.5	1.3	2.2	.9	.6	10.0
Globigerinoides ruber	1.7	0.2	0.2	.1	<.1	1.0
Globigerina quinqueloba	1.7	.5	1.2	.1	.1	4.8
Globigerinita glutinata	0.2					0.0
Globorotalia hirsuta				.1	.3	0.0
Globigerina calida				<.1	.2	0.0
Hastigerina pelagica				<.1	.1	0.0

from which to draw the output of empty shells, and hence a more rapid overturn is required. In view of the other uncertainties involved, a more detailed calculation on the basis of the distribution of sizes in the adult range seemed unjustified.

Eight plankton tows were taken with nets (150-μ mesh) during the nights of 29–30 and 30–31 August 1966, at the following depths: 15 to 0, 35 to 0, 45 to 35, 100 to 45, 300 to 100, 480 to 300, 500 to 300, and 450 to 0 m. Depths were controlled by an acoustic pinger attached to the nets, and the nets were closed by a tripping device operated by a propeller flowmeter. Between 250 and 500 m³ of water was filtered in each tow, and samples were preserved in buffered formalin. The foraminifera extracted by gravity separation from each sample were stained with methylene blue (5), which facilitated identification and recognition of live specimens. Half of the remaining part of each sample was washed and combusted (6), in order to obtain those shells that had not fallen out in the settling tube. Combustion at about 400°C preserves some carbon in living foraminifera. This is desirable because empty tests settle out preferentially in the gravity process, and thus falsify the ratios of live to dead foraminifera. Populations of living individuals that are larger than 150 μ are given in Table 1. From the multiple control afforded by the various overlapping tows, we conclude that the values are correct to within ± 50 percent.

The sediment trap is an umbrella-shaped free vehicle with an aperture of 0.7 m². The folded trap was released on 25 August 1966, and it sank to 6 m above the bottom within 2 hours. Approximately 6 hours later a Schick bimetal release parted and opened the trap. The instrument remained open for 4 days and then slowly pursed shut. A short time after the closing was completed a weight was released, and the trap floated to the surface. The contents were removed by pumping and subsequent washing of the trap. Results of the count of foraminifera larger than 150 μ (in this case 96 percent of the total) are tabulated in Table 1 (7).

It is necessary to evaluate whether the data are representative before interpretations are possible. The concentrations of foraminifera are probably reliable, since a similar series of tows taken 10 weeks later yielded comparable results. This suggests that the assumption of a steady state is valid. It is not possible, however, to be as confident about the sediment yield, since an attempt to repeat the experiment with a modified trap failed.

Sediment traps tend to increase the turbulence at the place of collection and this affects the results. It seems likely that the trap would catch less than the normal amount of tests settling over the area, since local turbulence would retard the sinking speed of the particles passing over the trap and thus increase their density at that position. It follows that fewer tests are then caught in the instrument. We believe, therefore, that the catch in the trap is a minimum amount. Bottom currents at the time of collection varied from 1 to 28 cm/sec, with an average current velocity of 6 cm/sec.

If the data are accepted with the reservations outlined, one can draw the following tentative conclusions: (i) The main growth of the species *Globigerina bulloides, Globigerina quinqueloba, Globigerinoides ruber,* and *Globoquadrina eggeri* is likely to take place in the upper layers of water, as shown by the relatively dense populations there. Also, the oxygen deficiency of the water at greater depths may be expected to be detrimental to growth and reproduction. (ii) The individuals below 100 m seem to have been brought in with submerged southerly water, judging from the assemblage of species there. Species that were not represented in the upper waters at the time did not contribute to the sediment catch on the bottom. They do not seem to reproduce at a rate comparable to the other species, and their turnover time must be very long under these circumstances. (iii) If the entire population below 100 m can be considered inert on the basis of these arguments, the turnover times for the species found in the surface layers are obtained in the following way. The densities given in Table 1 are converted to standing crops for the upper 100 m of the water column. These productive standing crops for each species are divided by the appropriate fluxes of empty tests, which are also given in Table 1. The resulting turnover times for *Globigerina bulloides, Globoquadrina eggeri, Globigerinoides ruber,* and *Globigerina quinqueloba* are 58, 33, 73, and 27 days, respectively (8). For reasons given above, the average life spans of these species should be shorter than the turnover times by a factor of 1 to 2.

These conclusions do not preclude the existence of a longer cycle, including the submergence of mature individuals under adverse environmental conditions. Such cycles are described for copepods (9) and may well run parallel to the shorter cycles of high productivity proposed here.

References and Notes

1. E. Boltovskoy, *Los Foraminiferos Recientes* (Eudeba, Buenos Aires, 1965), p. 204.
2. D. B. Ericson and G. Wollin, *Sci. Amer.* **207**, 1 (July 1962).
3. A. W. H. Bé and D. B. Ericson, *Ann. N.Y. Acad. Sci.* **109**, 1 (1963).
4. M. N. Bramlette, in *Oceanography*, M. Sears, Ed. (AAAS, Washington, D.C., 1961), p. 353.
5. W. H. Berger, *J. Paleontol.* **40**, 975 (1966).
6. K. N. Sachs, R. Cifelli, V. T. Bowen, *Deep-Sea Res.* **11**, 621 (1964). The authors use a temperature of 500°C.
7. The recovery of only large tests suggests that predation destroys the shells of ingested specimens.
8. Example calculation (*G. bulloides*): [(35 × 9.7) + (10. × 4) + (55 × 2.2)] m × number per m³/8.6 number per m² per day = 58 days.
9. J. E. G. Raymont, *Plankton and Productivity in the Oceans* (Pergamon, Oxford, 1963), p. 380.
10. Supported by the Marine Life Research Group at Scripps Institution of Oceanography and by the National Science Foundation. F. B. Phleger and J. D. Isaacs contributed interest, advice, and encouragement throughout this project. Basic assumptions underlying the calculations were clarified in discussions with E. W. Fager. R. Schwartzlose was scientist in charge during the cruise and provided the data on bottom currents. D. Brown and E. Duffrin assisted during the technical operations. Information in Fig. 1 was provided by the Data Processing Group, S.I.O.

BIOCHEMICAL STUDIES ON THE PRODUCTION
OF MARINE ZOOPLANKTON

By E. D. S. CORNER and C. B. COWEY

I. INTRODUCTION

A scientific approach to the study of mammalian nutrition began over 150 years ago with the development of analytical methods by the French school of chemists (e.g. Magendie, 1816). There is now a wealth of knowledge on the dietary needs and nutritional physiology of mammals. By contrast, the nutritional needs of invertebrates, and especially those of zooplankton, have been studied only for a decade or so.

The object of this review is to discuss experiments concerned with the nutrition of marine zooplankton. Information about the chemical composition of the animals and their diets is also included, but only in so far as it relates to the problem of secondary production in the sea; and data on respiration, together with rates of excretion of phosphorus and nitrogen, are dealt with as estimates of metabolic activity. In addition, particular attention is paid to the efficiencies of such vital processes as food assimilation, and growth. Particular metabolic pathways cannot be discussed as too little information is presently available.

II. THE NUTRITION OF ZOOPLANKTON

Because of the nature and aquatic habit of zooplankton, investigation of their nutrition is attended by greater experimental difficulties than are studies on mammals. Consequently there is a lack of precise information. The great advances in mammalian nutrition resulted from experiments in which chemically defined diets were fed and some parameter of growth was measured, or in which deficiency symptoms could be recognized when some dietary constituent was lacking. Few attempts have been made to repeat this type of experiment with aquatic invertebrates, but in view of the striking success of this approach when applied to salmon fingerlings by Halver and his collaborators (Halver, 1957; Halver, DeLong & Mertz, 1957) there seems no *a priori* reason why it should not be applied more widely.

(1) *Studies on zooplankton culture and growth*

Most experiments on the nutrition of zooplankton have been of a type designed to culture these animals on selected natural foods, i.e. species of phytoplankton. Where various algae are manifestly different in their ability to promote growth one might hope to make inferences of a sort regarding the nutrition of zooplankton. Different criteria have been used as an index of growth and thus of the food value of the algae used: for example, increase in length (Gibor, 1956), ability to support egg production (Marshall & Orr, 1952), ability to support several generations of zooplankton (Provasoli & Shiraishi, 1959; Provasoli, Shiraishi & Lance, 1959; Shiraishi & Provasoli, 1959). Some investigators have used wholly aseptic conditions; others have attempted to control bacteria with antibiotics; still others have largely ignored bacterial contaminants (if they may be so regarded). In general terms, the presence of bacteria in such experiments need not invalidate them: after all, bacteria-free animals were not used in the classical experiments on mammalian or fish nutrition, but no one seriously questions the findings on this score. In this connexion, Bernard (1961) observed that benthic as opposed to pelagic copepods may obtain nutriment from bacterial sources.

Particular pitfalls in experiments where algae are fed to zooplankton are: (i) the algal cells may be either too large or too small to be captured, ingested and used, (ii) the algae may have thick indigestible cell walls of cellulose, and (iii) the algae may secrete or excrete some noxious substance. Under any one of these conditions it is impossible to draw conclusions about the food value of the algal diet.

One of the earlier attempts to assay the food value of algal diets for zooplankton was that of Marshall & Orr (1952), who examined the numbers of eggs produced by *Calanus finmarchicus* fed on a variety of algae. Unfortunately, no efforts were made to see whether any of the eggs produced were viable. Starved animals were able to produce some eggs, but in markedly fewer numbers than fed animals and the authors were able to distinguish with reasonable clarity between 'effective' foods (*Ditylum brightwellii*, *Chlamydomonas* spp. and *Syracosphaera carterae*) and others (*Hemiselmis rufescens*) which did not raise the level of egg production above that of starving animals. An important observation of Marshall and Orr was that both *Hemiselmis* and *Dicrateria inornata* were utilized by *Calanus* (i.e. they were neither toxic nor indigestible) but neither was a good food as judged by the egg-laying criterion. It is reasonable to infer that these algae were deficient in some essential nutrients.

Marshall and Orr suggested that egg production (by ripe *Calanus*) be used to test the nutritive value of different food organisms because of its speed and simplicity. However, to be an acceptable test the eggs produced must be shown to be viable. Even then it is possible that certain constituents of the eggs are coming from other tissues of the animal (which are thereby depleted) rather than from the food. Some additional criterion than egg-laying is thus desirable in assaying nutritive value of foods: indeed it is to be doubted whether any experiment of short duration is truly acceptable as a measure of nutritive value.

Later, Gibor (1956) raised nauplii of the salt-water branchiopod *Artemia salina* from sterilized 'cysts', feeding them on algae collected from the brines of evaporation ponds, and using the length attained after 6 days in the culture medium as a measure of growth. He demonstrated that *Stephanoptera gracilis* was superior to other foods tested; *Dunaniella viridis* was a superior food to *D. salina*. Unfortunately *D. salina* appeared to vary in its food value in different experiments and it cannot be unequivocally stated that these closely related algae differ in food value for *Artemia*.

Walne (1956) compared the food value to oyster veligers of seventeen species of algae. He demonstrated that these algae varied highly in their value as food, growth being obtained with fourteen of the species tested but spatfall with only ten of them. In a similar study Davis & Guillard (1958) examined the relative values of different micro-organisms as foods for osyters and clam larvae. They found that *Isochrysis galbana* and *Monochrysis lutheri* were the best foods for oyster larvae, as judged by the mean body length achieved over a period of 12–14 days. They also made the important finding that mixtures of algal species such as *Monochrysis* together with *Platymonas* spp. and *Dunaniella euchlora* provided better growth than did any of these foods used singly. This means that any one of these algal foods was able to supplement either of the others; and each must therefore have contained nutrients in which the other two were wholly or (more probably) partially deficient. In further experiments using clam larvae, Davis and Guillard made a similar observation: a mixture of flagellates led to more growth by the animals than did equal quantities of any single species. Successful use of a mixture of flagellates, together with the diatom *Phaeodactylum tricornutum*, in rearing the pelagic copepod *Euterpina acutifrons* has recently been reported by Neunes and Pongolini (1965). Davis and Guillard believe that 'much of the difference in food value of closely related micro-organisms is directly attributable to differences in quantity or toxicity of their metabolites'. While the release of toxic metabolites may obviously influence the utilization of algae supplied as food it is difficult to see how the supplementary effects of different algae when fed in mixtures can be attributed to these external metabolites. It seems to us that the supplementing effect of one algal food upon another may be due to a relative deficiency of essential nutrients in different algal foods, this relative deficiency being made good when one alga is supplemented with another.

Urry (1965) assessed the food value of different algae for *Pseudocalanus elongatus* by measuring the survival time of this organism when it was fed on known concentrations of different algal cells under constant conditions. While the assay method left something to be desired in so far as there was no evidence that growth as opposed to maintenance was achieved on any food, and no algal food was able to maintain the animal indefinitely, the experiments were of long duration (up to 80 days) and distinct differences in food value of different algae were established. Of particular interest was the finding that two algae, *Dicrateria inornata* and *Nitzschia gotlandica*, had little nutritional value for the copepod. As both these organisms were ingested by the animal and as faecal pellets were produced it is not unreasonable to assume that both were non-toxic and acceptable as food. Their failure to maintain *P. elongatus* may be fairly ascribed to a nutritional inadequacy.

Provasoli *et al.* (1959) used bacteria-free algal diets to study the nutritional needs of *Artemia* and of the intertidal copepod *Tigriopus californicus*. They demonstrated that bacteria-free nauplii of *Artemia* reached adulthood if fed on one strain of *Dunaniella* but that another strain supported growth only to the IV metanauplius. Different strains of *Platymonas* gave similar effects. They also found that different food organisms supported different numbers of generations of *Tigriopus*, *Isochrysis galbana* giving rise to nine generations but *Syracosphaera elongata* only to one. Supplementary effects were again observed, *Rhodomonas lens* and *Isochrysis* being far better when used together than when used singly. These results were thought to be caused by differences in the chemical composition of the food organisms; and further circumstantial evidence in favour of this was provided by the finding that *Tigriopus* reproduced for only two generations when *Platymonas* was used as a food but could be cultured for 3 years on *Platymonas* supplemented by an unidentified bacterial flora.

Provasoli and his colleagues concluded that no single species of alga could support an indefinite number of copepod generations, because none of them supplied optimal amounts of all essential micronutrients.

Moyse (1963) has reported that barnacles from different thermal environments require different algal diets if they are to be successfully reared from nauplius to cyprid. Larvae of the arctic species *Balanus balanoides* required diatoms but not flagellates; those of the warm temperate species *Elminius modestus* were reared on both diatoms and flagellates; and those of the tropical species *Chthamalus stellatus* and *Lepas antifera* were reared on flagellates but not on diatoms. Whether these findings really represent some subtle effect of environmental temperature on dietary needs remains to be evaluated.

Evidence that animals may provide a better diet than algae as foods for some zooplankton has been reported by Lasker & Theilacker (1965), who were able to maintain *Euphausia pacifica*, *E. eximia* and *Nematoscelis difficilis* in the laboratory on a diet of *Artemia nauplii*. Compared with an algal diet consisting of *Platymonas subcordiformis* and *Dunaniella primulecta*, *Artemia* promoted a significantly faster rate of growth with virtually no mortality.

Mullin & Brooks (1967) have combined the diatoms *Cyclotella nana*, *Thalassiosira fluviatilis*, *Ditylum brightwellii* and animals (*Artemia nauplii*) in a diet to rear the copepods *C. helgolandicus* (pacificus) and *Rhincalanus nasutus* from egg to adult. They found, however, that not all the females maintained in culture copulated successfully; and those that did so failed to attain normal fecundity. Enriching the sea water with dissolved trace metals and organic growth factors did not increase fecundity. As a large number of animals survived it is possible that their diet was not grossly deficient but lacked some essential 'fertility substance' which was needed for gonad maturation and copulation. On the other hand this reproductive failure may not be a consequence of a nutritional deficiency. Other environmental and physiological factors may have been responsible: thus the particular laboratory conditions used (e.g. size of vessel, quantity and quality of light, temperature) may have militated against successful reproduction. Similar difficulties were encountered by Conover (1965), who reared *Calanus hyperboreus* from egg to adult but failed to obtain further generations because the animals would not copulate.

These various findings lead to the following conclusions. (1) Different algae vary markedly in their value as foods for zooplankton and some of these differences are a function of the chemical composition of the algae concerned. While some algae have been shown to be unacceptable because they secrete toxic substances, e.g. *Chlorella* (Lewin's isolate) (Loosanoff, Davis & Chanley, 1955) *Prymnesium parvum* (Davis & Guillard, 1958), others because of the nature of their cell wall, e.g. *Chlorella* (Cole, 1937), these reasons do not account for the observed inadequacies of *Hemiselmis rufescens, Dicrateria inornata* and *Nitzschia gotlandica* as diets (Marshall and Orr, 1952; 1955 a; Urry, 1965). (2) No single alga has so far been found that provides all the nutritional needs of any zooplankton animal; different algae supplement one another. (3) Some algae may not supply necessary growth factors for the elaboration of viable germ cells. (4) For some zooplankton, animal diets are superior to plant diets.

(2) Soluble growth factors

Provasoli and co-workers have begun to evaluate precise nutritional needs of zooplankton by using soluble growth factors to supplement algal diets. Thus, Shiraishi & Provasoli (1959) found that a mixture of B vitamins corrected the nutritional deficiencies involved in feeding *Isochrysis galbana* or *Chroomonas* spp. to *Tigriopus*. This animal seems to be affected by a lack of vitamins in three ways: larval development up to the first copepodite stage is retarded; the lifespan of the adult female is shortened; and the production of fertile eggs is inhibited. The experiments do not show whether the vitamins were assimilated by the algal food before its capture by the animals or whether *Tigriopus* assimilated them directly from the culture medium. The authors were also puzzled by the finding that they could replace their supplement of B vitamins by glutathione and still remove the nutritional deficiencies of the algal diet. Possibly this latter observation can be explained in terms of the relationship between glutathione and vitamin B_{12} (Ellenbogen, 1963), a compound known to be implicated among other things in the reduction of dithio groups in that its absence causes a fall in the levels of tissue–SH compounds, notably reduced glutathione. If this vitamin had been sufficient in itself to correct the nutritional deficiencies observed with the two species of algae (a possibility not tested by Shiraishi and Provasoli) then its replacement as a food supplement by glutathione would not be unexpected.

This type of approach to the problem of zooplankton nutrition promises to be of great use in identifying micronutrients essential to the animals; and it is interesting to note that the techniques may have wider application, as Lewis (1967) has now used them successfully to culture the early developmental stages of the copepod *Euchaeta japonica*.

In a further study, Provasoli & Shiraishi (1959) grew *Artemia* in axenic culture. The medium used was sea water containing trypticase, liver infusion, hydrolysed

102

RNA and DNA, horse serum, sucrose, cholesterol, Paramecium factor (extracted from autolysed yeast cells), glutathione, a mixture of B vitamins, and starch particles. They made the interesting finding that the liquid part of the medium containing all the nutrients did not support growth beyond the third-stage metanauplius; and that the starch particles, which promoted a swallowing action by the animals, were needed to ensure full development to the adult. They also found that thiamine and folic acid were needed (in addition to the quantities present in the liver extract and trypticase) as well as a source of sulphur-containing amino acids (either glutathione or cysteine), and horse serum (which could not be replaced by fatty acids and cholesterol). The essential need for starch particles to ensure adequate feeding suggests the possibility that marine herbivores may be able to absorb dissolved organic nutrients from the sea water, as long as they are actively ingesting particulate material (e.g. unicellular algae). This possibility is relevant to the important question of whether various organic micro-nutrients dissolved in sea water can act as food supplements for zooplankton. By and large, however, *Artemia* cannot utilize solutes readily, compared with—say—*Aedes* (Hutner & Provasoli, 1965).

In a more recent study (Provasoli & D'Agostino, 1962) *Artemia* were reared in aseptic culture, the medium containing mineral salts, amino acids, hydrolysed DNA and RNA, vitamins, sugars, fat-soluble factors and a fine particulate phase prepared from rice starch and heat-precipitated β- and γ-globulin and albumin. The experiments demonstrated that eight vitamins were necessary to produce sexually mature adults: thiamine, pyridoxamine, riboflavin, nicotinic acid, pantothenic acid, biotin, putrescine and folic acid. Removal of the thiamine arrested growth at the metanaupliar stage and lack of folic acid stopped the development of the young stages. The absence of other vitamins prevented growth and differentiation at stages between these two extremes. Finally, removal of carnitine, inositol and choline did not hamper growth, but these three vitamins were thought to participate in egg production because their absence reduced the number of ripe females.

The essential amino acid requirements of *Artemia* could not be identified by Provasoli and his colleagues because the animal does not utilize solutes efficiently. Perhaps this information might be obtained by an indirect approach, i.e. injecting the *Artemia* with a labelled carbon source ([^{14}C]glucose) and later isolating the tissue amino acids. Those amino acids into which ^{14}C is extensively incorporated can be synthesized by the animal and are not, therefore, essential dietary constituents. By inference the non-labelled tissue amino acids would represent essential dietary constituents. This method has been successfully applied to insect larvae (Kasting & McGinnis, 1958).

The investigations on *Artemia* indicate that crustacean nutrition is as complex as that of vertebrate nutrition and that the dietary requirements of Crustacea are rather similar to those of vertebrates. In at least some respects there appears to be a form of biochemical unity in animal nutrition—for instance, the amino acid requirements of those animals which have been investigated are remarkably similar (cf. Beerstecher, 1964).

III. CHEMICAL COMPOSITION OF ALGAL DIETS

In recent years attempts have been made to study the relationship between the chemical composition of various species of phytoplankton and their nutritional value. Thus, Parsons, Stephens & Strickland (1961) analysed eleven species for protein, carbohydrate and lipid; carbon, silicon, phosphorus and ash content; monosaccharides and amino acids in hydrolysates of whole cells. The bulk analyses showed that there was considerable variation in the relative proportions of protein, carbohydrate, and lipid, and the authors suggested that a ratio of 4:3:1 characterized a good diet because these proportions were found in *Monochrysis lutheri*, a species of good nutritional value for a variety of zooplankton (Walne, 1965; Urry, 1965). This is a very empirical approach and appears to overlook the need for vitamins and other growth factors by zooplankton: besides, *Phaeodactylum tricornutum*, which is a good food for *Pseudocalanus elongatus* (Urry, 1965), has a ratio of approximately 5:4:1, and *Skeletonema costatum*, which is regarded as a good food for *Calanus finmarchicus* (Marshall & Orr, 1952) has a ratio of approximately 7:4:1. Parsons *et al.* also suggested that the quantity of glucose occurring in each species might be a measure of nutritional value, as *Monochrysis lutheri* and *Skeletonema costatum* contained the highest percentage of carbohydrate that was readily hydrolysable and nutritionally useful. This suggestion has not yet been adequately tested, but no species of zooplankton has yet been shown to possess an absolute requirement for glucose. The carbohydrate contents of numerous species have been measured by various workers (Raymont & Conover, 1961; Raymont, Austin & Linford, 1964; Raymont & Linford, 1966) and found to be extremely small. These low values, however, do not rule out the possibility that carbohydrate may have a rapid rate of turnover in the animals and could still be of importance as an energy source.

Another carbohydrate found in certain diatoms and one which may be of importance to zooplankton is N-acetyl-D-glucosamine. McLachlan, McInnes & Falk (1965) have reported that this substance is present in large amounts in the mucilage of *Thalassiosira fluviatilis*, accounting for as much as 31–38 % of the total cellular material and being linked by β-(1 → 4) bonds to form the polymer termed 'chitan'. Later work (McLachlan & Craigie, 1966) has shown that the polymer is also present in *Cyclotella cryptica*. It would be interesting to know whether this polymer is widespread among members of the phytoplankton and whether it can be digested by zooplankton and the residues of N-acetyl-D-glucosamine used in the synthesis of chitin. Evidence of such a synthesis has so far only been obtained from *in vitro* studies with *Artemia* (Carey, 1965), but if the formation of chitin directly from dietary N-acetyl-D-glucosamine was possible in zooplankton—a process that in some insects normally seems to involve glucose, ATP, glutamine and acetyl coenzyme-A (Candy & Kilby, 1962)—then it obviously has an important nutritional use.

(1) *Proteins and amino acids*

Parsons *et al.* (1961) reported that the relative proportions of amino acids in acid-hydrolysates of various species of phytoplankton showed some marked differences.

For example, the ratio of aspartic acid to glutamic acid was roughly 4:1 in *Monochrysis lutheri*, 2:1 in *Skeletonema costatum* and 1:1 in *Phaeodactylum tricornutum;* again, whereas *Monochrysis* contained twice as much lysine as glycine, the opposite was found with *Phaeodactylum*, and the lysine content of *Skeletonema* was only detected, not estimated. Such large differences in amino acid composition—implying similar differences in nutritional value—were not observed by Cowey & Corner (1963*a*) in analyses of particulate material collected at all seasons from the coastal waters off Plymouth and these authors concluded that variation in amino acid composition was probably not a factor limiting zooplankton production, at least in this particular sea area. Nevertheless, because of the apparent contradiction between the two sets of data, further analyses were made of several algal species grown under optimal conditions and kept bacteria-free (Cowey & Corner, 1966). These data are included in Table 1 and it will be seen that, contrary to the findings of Parsons *et al.*, there is little variation among species in terms of amino acid composition. In a more recent study by Chau, Chuecas & Riley (1967), gas–liquid and thin-layer chromatography were used to estimate twenty-five amino acids in twenty-five species of phytoplankton. Data for six of these species, three of high and three of low nutritional value, are also shown in Table 1. There is very fair agreement between the data of Cowey and Corner and those obtained by Chau *et al.* for the same species: in addition it is interesting to note that the amino acid compositions of species with high nutritional values (*Isochrysis galbana, Monochrysis lutheri* and *Phaeodactylum tricornutum*) closely approximate to those of species with low values (*Cricosphaera elongata, Dicrateria inornata* and *Hemiselmis* spp.). This finding is borne out by some data of Walne (Walne, 1968, personal communication) as yet unpublished but which he has kindly allowed us to quote. These data (Table 2) comprise amino acid analyses of 'good' (*Tetraselmis suecica*) and 'poor' (*Chlamydomonas coccoides*) foods for juvenile clams (*Mercenaria mercenaria*). The proteins of the two foods are very similar in amino acid composition and approximate closely in over-all composition to those of the clam.

Further information on the nutritional value of algal proteins has been obtained by Ogino (1963). Using data by Halver *et al.* (1957) for amino acids indispensable to fish, he applied the method of Oser (1959) in order to calculate the essential amino acid (EAA) index of three species of diatom. The values for *Skeletonema costatum* (70·1), *Chaetoceros simplex* (80·5) and *Amphipora* spp. (77·8) were relatively high and compared well with those for animal protein extracted from *Daphnia* spp. (84·7) and *Chironomus plumosus* (90·0).

At the moment, it seems probable that the nutritional value of marine algal proteins are generally high and that differences between the dietary values of various species are not related to differences in amino acid composition. However, before this question can be finally resolved it will be necessary to know more about the biological availability of these amino acids. Thus, not all the amino acids in a protein are biologically available to the fed animal. In some cases they may be combined with other compounds and thereby made inaccessible: for example, the ϵ-amino group of lysine can be conjugated with a sugar residue in a Maillard reaction. There is a method by which

105

the biological values of (phytoplanktonic) proteins can be assayed rapidly by using the actively proteolytic organism *Streptococcus zymogenes* (Ford, 1960). Its use, of course, invites the criticism that one is measuring the biological value of protein for *S. zymogenes* and not for zooplankton. However, there is a basic similarity in pattern of animal nutrition (see above) and the use of the *S. zymogenes* assay suggests itself as the best available method of measuring biological value of phytoplanktonic protein and relating it to amino acid composition.

(2) *Fatty acids*

The first detailed analysis of the fatty acids in a species of marine phytoplankton was made by Lovern (1936) using *Nitzschia closterium*. He found that the main constituents were C_{16} and C_{18} unsaturated acids, together with a C_{16} fully saturated

Table 1. *Comparison of amino acid compositions of 'good' and 'poor' algal foods*

(All values expressed as g. amino acid/100 g. amino acids.)

test animal ...	*Tigriopus californicus**			*Pseudocalanus elongatus†*			*Ostrea edulis‡*		
	'good'	'poor'		'good'	'poor'		'good'	'poor'	
	Iso-				*Dicra-*			*Hemi-*	
	chrysis	*Cricosphaera*		*Phaeodactylum*	*teria*		*Monochrysis*	*selmis*	
algal species ...	*galbana*	*elongata*		*tricornutum*	*inornata*		*lutheri*	spp.	
amino acid	(a)	(a)	(b)	(a)	(b)	(a)	(a)	(b)	(a)
alanine	9·7	10·1	6·7	11·3	7·0	8·9	10·5	8·4	9·4
mono-AA	—	0·3	—	0·8	—	2·0	0·4	—	0·7
di-AA	—	0·2	—	0·2	—	0·5	0·2	—	0·2
2AIB	0·5	0·6	—	0·6	—	0·7	0·5	—	0·4
2ANB	1·3	1·5	—	1·8	—	1·4	0·7	—	1·6
4ANB	—	—	—	—	—	—	—	—	0·5
arginine	5·7	5·0	6·1	4·5	5·3	6·6	7·7	5·7	7·2
aspartic acid	9·9	9·6	11·4	9·9	10·5	9·1	9·6	9·3	9·5
cysteine	0·5	—	—	—	1·5	—	—	1·4	0·9
glutamic acid	8·4	12·1	11·1	14·1	13·1	7·5	10·8	10·6	8·3
glycine	6·3	5·7	5·5	5·6	5·6	6·0	4·9	5·8	6·1
histidine	1·9	1·8	2·2	0·7	1·9	1·6	1·7	2·1	2·6
iso-leucine	3·3	4·6	3·5	4·5	4·9	4·0	4·6	4·4	4·7
leucine	10·2	9·4	9·4	7·5	8·3	9·9	9·0	10·0	9·6
lysine	7·3	5·7	5·5	4·9	6·8	6·9	7·2	6·8	8·0
methionine	3·2	3·2	4·4	2·2	2·1	3·8	3·2	2·7	3·1
ornithine	0·4	—	—	—	—	0·3	0·5	—	0·5
phenylalanine	4·4	7·1	6·4	8·5	5·6	4·4	6·5	5·3	3·8
proline	6·7	5·6	4·9	4·9	6·0	5·7	5·3	4·2	4·7
serine	6·0	4·7	4·4	3·4	5·3	6·3	3·5	5·2	6·8
threonine	5·0	4·5	5·0	5·8	4·9	5·8	4·7	4·9	5·1
tryptophan	0·4	—	1·3	—	—	—	—	2·0	—
tyrosine	2·1	3·2	6·1	3·7	3·7	2·2	3·4	4·5	1·5
valine	6·8	5·1	6·1	5·1	7·5	6·4	3·7	6·7	4·8

(a) Values from Chau, Chuecas & Riley, 1967; (b) values recalculated from Cowey & Corner, 1966. Abbreviations of amino acid names: mono-AA = 2-amino-adipic acid; di-AA = 2,5-diamino adipic acid; 2AIB = 2-amino-*iso*-butyric acid; 2ANB = amino-*n*-butyric acid; 4ANB = 4-amino-*n*-butyric acid.

* Provasoli, Shiraishi & Lance, 1959,   † Urry, 1965,   ‡ Walne, 1965.

acid (palmitic). Farkas & Herodek (1964) measured the fatty acids in two freshwater species, *Scenedesmus obusiusculus* and *Chlorella pyrenoidosa*: after hydrogenation the main constituents of *Scenedesmus* were a C_{16} acid and a C_{18} acid, the former accounting for 47·4 % and the latter for 52·6 % of the total content of fatty acids: corresponding values for *Chlorella* were 35·7 and 64·3 respectively. Kayama, Tsuchiya & Mead (1963) used gas–liquid chromatography to examine the fatty acids in the marine diatom *Chaetoceros simplex* and found that the main constituents were the fully saturated C_{14} (myristic) and C_{16} (palmitic) acids—which accounted for 13 % and 18·1 % of the total respectively—together with the C_{16} unsaturated palmitoleic acid (47·9 %) and the C_{18} unsaturated oleic acid (8·7 %). Small amounts of linoleic and linolenic acids were also present.

Table 2. *Comparison of the amino acid composition of 'good'* (Tetraselmis suecica) *and 'poor'* (Chlamydomonas coccoides) *foods and of juvenile clams* (Mercenaria mercenaria) *fed on them**

(Molar ratios relative to leucine ($= 1·00$).)

	Tetraselmis	Chlamydomonas	clams fed on Tetraselmis	clams fed on Chlamydomonas
cystine/2	0·14	0·10	0·22	0·25
methionine	0·30	0·25	0·33	0·35
aspartic acid	1·07	1·28	1·68	1·57
threonine	0·74	0·68	0·93	0·92
serine	0·70	0·69	1·09	0·92
glutamic acid	1·11	1·20	1·62	1·68
proline	0·40	0·63	0·57	0·59
glycine	1·21	1·13	1·10	1·07
alanine	1·51	1·31	1·05	1·04
valine	0·76	0·74	0·82	0·77
isoleucine	0·49	0·47	0·65	0·71
leucine	1·00	1·00	1·00	1·00
tyrosine	0·26	·026	0·30	0·36
phenylalanine	0·47	0·45	0·41	0·47
lysine	0·69	0·74	0·92	0·98
histidine	0·20	0·30	0·44	0·39
arginine	0·50	0·40	0·65	0·84

* Walne & Youngson (1968) unpublished data.

A study of the fatty-acid compositions of various marine unicellular algae has recently been completed by Ackman, Tocher & McLachlan (1968). These are summarized in Table 3, from which it is clear that striking differences occur in the relative proportions of fatty acids found in the various species. For example, *Amphidinium carteri* and *Syracosphaera carterae* have significant amounts of a C_{22} unsaturated acid which is found only in traces in the other six species; 50 % of the total fatty-acid content of *Cyclotella cryptica* is in the form of C_{16} unsaturated acids only traces of which are found in *Amphidinium*; unsaturated C_{18} acids account for 54 % of the total fatty-acid content of *Dunaniella tertiolecta* but the corresponding value for *Cyclotella cryptica* is only 4 %. A surprising feature of the data is the presence of C_{22} acids not found by Lovern and by Kayama *et al.*: moreover, all the species studied by Ackman *et al.* were in a mature stage of growth and, in a previous study (Ackman,

Jangaard, Hoyle & Brockerhoff, 1964) with *Skeletonema costatum*, it was shown that most polyunsaturated C_{20} and C_{22} acids decreased as the culture aged (see Table 4). The fatty-acid composition of an alga can also change in response to other factors; for example, the level of available nitrate (Mangold & Schlenk 1957).

There are, then, striking variations in the fatty-acid composition of various unicellular algae. An important point may be that most algae have a relatively high content of polyunsaturated fatty acids. In some the predominant unsaturated acid is of chain length C_{16}, in others C_{18}, C_{20} or even C_{22}; that is, the chain length may vary but polyunsaturated acids are present in relatively high amount in all species examined.

In mammals there is increasing evidence to suggest that pathological conditions may occur when there is a relative deficiency in polyunsaturated fatty acids in the diet: whether or not zooplankton have a similar essential fatty-acid requirement is not known but the implication is that any such requirement would be met by almost any algal species which has yet been examined.

Table 3. *Fatty acids in various species of phytoplankton*

(All values as percentages of total weight of fatty acids.)

species	saturated acids			unsaturated acids				
	C_{14}	C_{16}	C_{18}	C_{14}	C_{16}	C_{18}	C_{20}	C_{22}
Cyclotella cryptica	7	28	tr	tr	50	4	8	tr
Syracosphaera carterae	1	22	tr	tr	7	53	4	10
Olisthodiscus spp.	8	22	tr	1	23	17	22	3
Amphidinium carteri	3	36	6	tr	1	18	8	26
unknown cryptomonad	1	15	1	tr	2	61	16	tr
Dunaniella tertiolecta	1	16	1	tr	25	54	tr	—
Tetraselmis spp.	1	20	tr	tr	20	46	10	1
Porphyridium spp.	1	33	3	tr	4	10	45	tr
*Nitzschia closterium**	8	17	2	1	36	20	16	—

* Data from Lovern (1936).

Less than 1% either omitted or trace (tr).

Table 4. *Change in fatty-acid composition of* Skeletonema costatum *with age of culture*

(All values as percentages of total weight of fatty acids.)

acid	days of growth				
	2	4	6	8	10
C_{14} saturated	9·6	26·5	30·1	31·8	32·9
C_{16} saturated	9·0	6·1	6·6	7·1	6·8
C_{18} saturated	1·0	tr	tr	tr	0·2
C_{14} unsaturated	0·9	1·8	3·1	4·5	3·0
C_{16} unsaturated	37·0	40·6	37·4	35·7	35·9
C_{18} unsaturated	9·2	5·6	4·8	4·1	4·8
C_{20} unsaturated	25·6	17·7	16·4	13·5	14·1
C_{22} unsaturated	7·1	1·3	0·8	1·0	1·2

tr = trace.

(3) Vitamins

In view of the obvious need by zooplankton for a dietary source of vitamins (Provasoli & Shiraishi, 1959; Provasoli & D'Agostino, 1962) it would be interesting to know the amounts of these substances available in different algal species. So far this kind of study seems to have been confined to freshwater algae (Morimura, 1959) and the three species examined were found to differ markedly in their vitamin content. The possibility that this variation is also characteristic of marine species and may reflect their individual nutritional values as foods for marine zooplankton seems not to have been examined. Some data have been reported, however, for the quantity of vitamin B_{12} in phytoplankton from coastal waters and from oceanic waters (Cowey, 1956). The value obtained was 0·02–0·03 μg./g. wet weight (as compared with 0·06–0·07 μg./g. wet weight for zooplankton collected from the same sea areas). Whether the animals obtain most of their supplies of this vitamin from algal diets, or from bacterial flora inhabiting their guts, is still an open question.

IV. CHEMICAL COMPOSITION OF ZOOPLANKTON

The previous section dealt with the chemical composition of various algal diets used by zooplankton, and particular attention was given to amino acids and fatty acids, mainly because these represent the only groups of substances which have been studied in any detail. In this section the amino acid and fatty-acid compositions of phytoplanktonic algae are compared with those of zooplanktonic species, the point being whether assimilated foodstuffs are to a large extent incorporated unchanged into animal tissues or whether they need to be 'edited' by the animals. Such considerations are relevant to the question of nutrition, for the more metabolic interconversion which is necessary the greater is expenditure of energy. The extent to which such metabolic interconversion occurs must influence the over-all efficiency of the feeding process.

(1) Proteins and amino acids

Amino acids, either in the free state or combined as peptides and proteins, account for a very high proportion of the total organic nitrogen in zooplankton. Thus, Christomanos, Dimitriadis and Gardiki (1962) have shown that in mixed zooplankton (mainly *Centropages lamatus*, *Oithona similis*, *Temora longicornis* and *Acartia bifilosa*) 70–80 % of the total nitrogen is present as amino acids; Cowey & Corner (1963 b) found that 90 % of the total nitrogen in *Calanus finmarchicus* was amino acid nitrogen (76 % as protein and 14 % as free amino acids); and in a parallel study (Cowey & Corner, 1963 a), carried out during most of the year, these authors found that amino acids in *Calanus helgolandicus* accounted for roughly 80 % of the total nitrogen. In addition, virtually all the organic nitrogen in *Neomysis integer* is present as protein (Raymont, Austin & Linford, 1964), and this is so at all seasons (Raymont, Austin & Linford, 1966 a). Furthermore, 75–80 % of the total organic nitrogen is accounted for as protein in the planktonic decapods *Acartia phyra pupurea*, *Gennadas clavicarpus* and *Sergestes* spp. (Raymont, Austin & Linford, 1966 b).

It can be inferred from these analyses that there is little chitin present in these zooplanktonic animals and thus that little of the dietary protein nitrogen is deflected from protein synthesis to the synthesis of N-acetyl glucosamine. Furthermore, although urea, trimethylamine oxide and similar compounds are present in relatively large quantities in some aquatic animals this does not seem to be so with zooplankton.

The first complete analyses of the amino acid composition of zooplanktonic animals was made with *Calanus helgolandicus* (Cowey & Corner, 1962) and *C. finmarchicus* (Cowey & Corner, 1963 b). In a further study data were obtained for *C. helgolandicus* at most seasons (Cowey & Corner, 1963 a) as well as for particulate material (mainly phytoplankton) collected from the same sea area. A summary of the results is shown in Table 5 (taken from Corner & Cowey, 1964) and it is interesting to note that average values (March–December) for *C. helgolandicus* are similar to those for the particulate material on which the animals feed, the only exceptions being

Table 5. *Amino acid composition of adult female* Calanus helgolandicus *and of particulate material from Plymouth Sound*

(Values are the means of nine determinations made throughout the year. Figures in parentheses show the range of values encountered.)

amino acid	g. amino acid N/100 g. total amino acid N		amino acid	g. amino acid N/100 g. total amino acid N	
	Calanus	particulate material		*Calanus*	particulate material
arginine	17·9 (15·3–20·9)	15·9 (11·1–22·7)	threonine	3·9 (3·6–4·3)	4·6 (3·2–5·3)
glycine	11·9 (9·1–13·7)	14·3 (11·4–17·7)	proline	3·9 (2·9–5·0)	3·6 (1·9–5·2)
lysine	11·2 (9·5–13·8)	9·3 (6·6–10·8)	iso-leucine	3·7 (3·5–4·3)	4·3 (3·4–5·3)
alanine	9·1 (7·3–12·0)	7·3 (6·1–9·3)	histidine	3·6 (2·8–4·8)	2·2 (0·9–3·4)
glutamic acid	8·1 (6·8–9·5)	7·7 (4·6–10·5)	tyrosine	2·5 (1·6–3·4)	1·9 (1·3–2·1)
aspartic acid	7·1 (6·2–8·4)	7·7 (4·6–9·5)	phenylalanine	2·4 (1·8–2·9)	2·7 (1·5 3·8)
leucine	6·0 (5·4–6·6)	6·3 (5·9–7·3)	taurine	1·7 (1·2–2·8)	—
valine	5·7 (4·5–8·2)	5·7 (5·0–6·5)	methionine	1·2 (0·8–1·8)	0·8 (0·1–1·5)
serine	3·9 (3·4–4·7)	8·2 (6·1–9·9)	cystine/2	0·6 (0·3–1·0)	0·11(0·10–0·12)

that the diet contains no taurine, but a relatively high concentration of serine. It therefore follows that apart from their need to synthesize taurine (presumably from sulphur-containing amino acids in the diet) the animals have little to do by way of changing the relative proportions of the dietary amino acids, which are already in approximately the right amounts for incorporation into *Calanus* protein. Protein synthesis will therefore be facilitated in these animals as long as these same relative proportions of amino acids are maintained after absorption and transport to the sites of protein synthesis. Evidence consistent with the view that the amino acids are still in the same relative quantities after absorption has been provided by Cowey & Corner (1966), who showed that the proportions of various amino acids in the faecal pellets produced by *C. finmarchicus* were very similar to those in the plant (*Skeletonema costatum*) on which the animal fed.

The only other detailed study of the amino acid composition of herbivorous zooplankton is that by Barnes & Evens (1967), who examined eggs of *Balanus balanoides* and *B. balanus*. The relative amounts of amino acids were close to those observed by Cowey & Corner (1963b) for *Calanus finmarchicus* and—as would be expected—the resemblance was closest when later stages of the eggs, containing the fully developed embryos, were considered.

(2) Fatty acids

Compared with proteins, lipids generally account for a smaller percentage of the dry body weight of zooplankton (Baalsrud, 1955; Schwimmer & Schwimmer, 1955). Individual values vary considerably from species to species (Yamada, 1964) and for any one species the lipid content changes with stage of development (Orr, 1934); for related genera it varies with geographical distribution (Littlepage, 1964); and in some cases it also reflects the availability of food (Conover, 1962; Littlepage, 1964).

Table 6. *Fatty acids of zooplankton oils*

(All values as percentage of total weight of fatty acids.)

	saturated acids					unsaturated acids						
	C_{12}	C_{14}	C_{16}	C_{18}	C_{20}	C_{12}	C_{14}	C_{16}	C_{18}	C_{20}	C_{22}	C_{24}
Calanus finmarchicus	—	12·3	8·9	4·0	—	—	0·3	11·2	15·6	20·9	25·2	1·6
Euphausia superba	0·5	11·9	14·4	1·4	—	0·4	4·5	18·6	34·7	(———13·6———)		
E. pacifica	—	1·0	13·8	1·3	—	—	7·0	8·6	52·8	(———15·5———)		
Neomysis nakazawai	—	5·4	13·9	3·4	—	—	1·3	8·9	15·8	33·5	17·5	—
N. intermedia	—	4·5	24·5	11·5	1·5	—	1·0	11·0	18·0	(———28·0———)		
Calanus plumchrus	—	19·3	15·0	2·1	—	—	0·3	4·9	18·2	15·2	24·9	—
Parathemisto japonica	—	7·0	12·5	2·5	—	—	1·0	5·0	11·0	37·0	25·0	—

Studies on the nutrition of zooplankton in terms of lipids have been mainly confined to the fatty-acid fraction, and in contrast to what has been found in studies of amino acids, the animals seem able to modify considerably the fatty acids in algal diets. Thus, Lovern (1935) observed that *C. finmarchicus* contained a high percentage of C_{20} and C_{22} unsaturated acids which were not present in *Nitzschia closterium*. Farkas & Herodek (1964) confirmed this observation, but found that much lower proportions of these acids were present in a warm-water species of copepod, namely *Paracalanus parvus* from the Bay of Naples.

Further evidence that zooplankton elongate dietary fatty acids was reported by Kayama, Tsuchiya & Mead (1963), who showed that the fatty acids in *Artemia salina* fed on *Chaetoceros simplex* contained a significant fraction (12 %) of the C_{20} eicosapentaenoic acid which was absent from the plant. However, the relative proportions of fatty acids in zooplankton can vary considerably with species, as is shown by the work of Yamada (1964) summarized in Table 6. The results confirm earlier findings that C_{20} and C_{22} acids account for most of the fatty-acid content of *Calanus finmarchicus*, and show that this is also the case with *C. plumchrus* and two species of mysid. These longer-chain fatty acids, however, are present to a much lesser extent in euphausiids and so the degree of elongation of dietary fatty acids seems to vary considerably with species. It is also interesting to note that, contrary to the findings of Farkas & Herodek (1964) with *Paracalanus parvus*, the cold-water species *Euphausia superba* has a relatively low percentage of longer-chain fatty acids. Possibly the patterns of fatty acids in zooplanktonic animals reflect their feeding habits, relatively large amounts of the longer-chain fatty acids being characteristic of herbivores such as *C. finmarchicus*, but less so of animals that can also be carnivorous, such as *E. superba*.

It was pointed out earlier that some higher animals have a dietary requirement for polyunsaturated fatty acids. The ability of certain zooplankton to synthesize C_{20} and C_{22} unsaturated fatty acids is interesting and points to a biochemical diversity between the synthetic abilities of these creatures and those of mammals. In those species where this capacity to synthesize C_{20} and C_{22} unsaturated acids is found it is perhaps unlikely that a dietary requirement for unsaturated acids exists.

(3) *Hydrocarbons*

A compound of some interest in zooplankton is the hydrocarbon pristane (2,6,10,14-tetramethylpentadecane), concentrations amounting to 1–3 % of the total lipid having been extracted from three species of copepod by Blumer, Mullin & Thomas (1963 a). The same authors (Blumer, Mullin & Thomas 1963 b) described the formation of pristane from phytol and pointed out that its rate of accumulation depends on the quantity of chlorophyll ingested: the pristane level in the animals might therefore serve as an 'indicator' of the amount of phytoplankton assimilated and provide useful data for productivity studies, particularly those carried out at sea. The specific gravity of the hydrocarbon is 0·78 whereas the average value for other lipids is 0·88: thus pristane may also be concerned in buoyancy regulation during periods of starvation. In buoyancy terms the retention of 1 μg. of pristane by the animal releases 1·6 μg. of other lipids for metabolism. It should be pointed out, however, that the highest concentration of pristane found was in *Calanus hyperboreus* where it accounted for only about 3 % of the total lipids (and less than 1 % of the dry body weight of the animal). This means that even with its low specific gravity pristane only gives about 5 % of the lift provided by these other lipids.

V. METABOLIC ACTIVITY

An important factor influencing the over-all efficiency with which zooplankton make use of captured food is the rate at which various body constituents are metabolized and need to be replaced by foodstuffs which would otherwise be used for growth and reproduction. Numerous studies have been made of the metabolic activity of different species of zooplankton and the methods most often used have involved either measuring oxygen uptake or the excretion of nitrogen or phosphorus. The present section deals with recent investigations of this problem.

(1) *Excretion of phosphorus*

Many measurements have been made of the release of phosphorus into the surrounding water by zooplankton. Generally very high values have been obtained and these have sometimes been assumed to indicate very high metabolic activity on the part of the animals. Thus, Harris (1959) found that mixed zooplankton (mainly *Acartia tonsa* and *A. clausi*) from Long Island Sound released 11·47 μg. P/mg. dry body wt./day

112

into the water. This value, together with the observation by Harris & Riley (1956) that the phosphorus content of the zooplankton was 0·82 % of their body weight, can be used to show that Harris's animals liberated phosphorus equivalent to nearly 150 % of their total body phosphorus each day. Similarly Beers (1964) showed that the warm-water chaetognath *Sagitta hispida* liberated 2·39 µg. P/mg. dry wt./day, which was equivalent to nearly 40 % of its total body phosphorus. Pomeroy, Mathews & Min (1963) found that the daily quantity of phosphorus released from warm-water zooplankton represented the equivalent of about 100 % of their body phosphorus, and showed by differential analysis that a large fraction of this released material was in the form of organic phosphorous compounds. They concluded that a high rate of release was typical of small, metabolically active zooplankton. This view has been expressed in more general terms by Johannes (1964 a): he suggested that the time taken by a marine animal to release an amount of dissolved phosphorus equivalent to its total phosphorus content is a function of its body size. From his data it may be shown that the very small ciliate *Euplotes vannus* (dry body wt. *ca.* $1·0 \times 10^{-3}$ µg.) liberates the equivalent of its body phosphorus in about 30 min.; whereas the corresponding figure for a larger animal such as *Calanus finmarchicus* (dry body wt. *ca.* 0·3 mg.) is roughly 20 hr.

The values for phosphorus release by zooplankton are undoubtedly of great significance from the ecological standpoint, the return of phosphorus to the water being vital to the growth of phytoplankton (Ketchum, 1962). They cannot yet, however, be used as an index of the metabolic activity of the zooplankton because there is no unequivocal evidence to show that all the phosphorus released was ever actually assimilated by the animal; that is, that it was ever wholly incorporated into the tissues. The release of soluble compounds by an aquatic animal cannot be taken to imply that these products arise from material metabolized by the animal. These measurements, then, do not represent true excretion by the animal, unless they are accompanied by additional evidence to show that the substances being measured do not arise from non-assimilated food.

Odum (1961), using ^{65}Zn as a metabolic indicator, pointed out that the rate of release of ingested (radioactive) substances takes place in two stages. First, non-assimilated food as well as metabolic waste products are released; then, in the second stage, metabolic wastes alone are released. The latter stage represents excretion. Bearing this in mind, the experiments of Marshall & Orr (1961) in which the rate of excretion of phosphorus by *C. finmarchicus* was estimated by examining the rate of loss of ^{32}P from animals that had been fed [^{32}P]-labelled algae, probably represent a more accurate value for phosphorous excretion. They found that the animals excreted the equivalent of their body phosphorus in about ten days (cf. Johannes, 1964 a).

The differentiation between 'excreted' and 'liberated' phosphorus has been made clearly by Johannes (1964 b). From his experiments using [^{32}P]-labelled algae and dissolved ^{32}P compounds he was able to calculate two turnover times, namely the physiological turnover time (i.e. the time for an amount of phosphorus equal to that in the tissues of the animal to pass through these tissues) and an ecological turnover time (i.e. the time for an amount of phosphorus equal to that in the tissues to pass through an animal irrespective of whether or not it is assimilated). The physiological turnover time for the amphipod *Lembos intermedius* was 41 hr. The ecological turnover time was only 6·6 hr.

It has been pointed out (Redfield, Ketchum & Richards, 1963) that complete oxidation of organic material with composition similar to that of plankton (C:N:P = 106:16:1) requires 276 atoms of oxygen per atom of phosphorus. Experimental determinations of the O:P ratio in plankton have given results only $\frac{1}{5}-\frac{1}{4}$ of this value (Harris, 1959; Satomi & Pomeroy, 1965). The probable explanation of these low values in our view is that the phosphorus excretion figures do not only represent phosphorus metabolized and excreted by the animals, they also include phosphorus which has never been assimilated. Satomi & Pomeroy (1965), however, hold that a large fraction of the phosphorus is excreted in organic form. They draw attention to the possibilities that (1) P—O—P bonds may be hydrolysed more rapidly than others and thereby speed the flux of phosphorus, (2) zooplankton may engage in some form of aerobic fermentation and excrete lactic acid or other organic acids and thereby consume less oxygen, (3) the food of zooplankton may not be as rich in phosphorus as is the zooplankton itself. It is not clear to the present reviewers how the last possibility explains the low O:P ratios found.

On the basis of Marshall & Orr's (1961) data Conover (in an addendum to their paper) has postulated the existence of at least two pools of phosphorus in *Calanus finmarchicus*. One of these pools is labile, with a half-life of 0·375 days, and constitutes up to 6 % of the total phosphorus; the other pool is more stable, with a half-life of 13 days, and comprises 94–99 % the total phosphorus. It is still necessary to identify chemically these phosphorus pools in terms of particular phosphate fraction(s) within the animal; and it must be recognized that permeability barriers probably exist between similar phosphorus compounds in different tissues. For example, in the absence of a rapid circulatory exchange between different tissues it is difficult to see how the phospholipid in, say, the ovaries, nerve cord and gastric-tract walls can form a uniform pool.

It follows that before the concept of phosphorus pools can be advanced, information is necessary on the different chemical forms of phosphorus present in zooplankton (i.e acid-soluble nucleotides, phospholipid, nucleic-acid phosphorus, and phosphoprotein) and on the lability of each. The techniques available for obtaining the basic chemical data have been critically reviewed by Hutchison & Munro (1961).

(2) *Oxygen consumption*

Many studies have been made of the respiration of various species of zooplankton, and much of the earlier work has been summarized by Marshall & Orr (1961). The methods used have varied from Warburg manometry (Raymont & Gauld, 1951; Vacelet, 1961) to polarography (Teal & Halcrow, 1962), but most of the data have been obtained with the chemical method of Winkler (*see* Conover, 1956). From a biochemical viewpoint it is useful to be able to evaluate oxygen consumption in terms of quantities of body constituents metabolized. This, however, is difficult to do because no direct measure of respiratory quotient has been made for zooplankton.

Thus, Marshall & Orr (1958) obtained values for the respiration rate of ripe female *Calanus finmarchicus* at 10° C.: they equated the values for oxygen consumed with the fraction of dry body weight used daily and found that this was 3·9–7·2 % in summer and 2·8–6·7 % in winter, depending on whether carbohydrate (high values) or lipid (low values) was used as a substrate. In some later studies measurements of respiration rate have been complemented by chemical analyses of changes in the levels of lipid in the animals and evidence has been obtained that the metabolism of lipid predominates in certain species when starved. Thus, Conover (1964) reports that lipid accounted for most of the body weight lost by *C. hyperboreus* when starved for some 14 weeks; and Littlepage (1964) observed that the lipid content of the Arctic species *Euphausia crystallorophus*, which accounted for 35–36 % of the dry body weight, decreased at a steady rate throughout the winter, when food was scarce, to a minimum of 9·4 % just before the major phytoplankton bloom.

Table 7. *Respiration rates and theoretical amounts of lipid metabolized by certain species of zooplankton*

(Respiration rate of *A. clausi* from Conover (1960) using July values; lipid content from Yamada (1964). Respiration rates and lipid contents of *M. longa*, *C. finmarchicus*, *C. hyperboreus* and *P. norvegica* from Conover & Corner (1968). Respiration of *E. superba* from Lasker (1966) and lipid content from Fagerlund (1962).)

species	dry body wt. (mg.)	respiration rate (μl. O_2/mg. dry wt./day)	lipid content (μg./mg. dry body wt.)	% lipid used daily
Acartia clausi	0·0045	68·9	58	57·5
Metridia longa	0·25	35·0	160	10·5
Calanus finmarchicus	0·40	25·0	250	4·9
C. hyperboreus	2·30	12·5	350	1·7
Pareuchaeta norvegica	3·0	17·5	300	2·8
Euphausia superba	5·4	35·3	98	17·4

The data in Table 7 have been used to illustrate the proportion of body lipids which would be metabolized daily by certain species of zooplankton, assuming that lipid was the only substrate. The data show that in spite of differences of experimental method and temperature, as well as of sea area and season, there is adequate evidence that the oxygen consumption of copepods in terms of the body weight falls markedly as body weight increases; so, too, does the theoretical percentage of body lipid used in metabolism. By contrast, the respiration rate of the euphausiid is high, as is its calculated daily expenditure of lipid. The high turnover rate of lipid in the small copepod *Acartia clausi* (based on data for oxygen consumption) implies that these animals must impound a large quantity of food daily in order to survive: and recent data by Petipa (1966) show that the food requirement of this species, calculated from values for respiration rate, varies between 46·2 % (adult) and 106·5 % (nauplius) of the fresh body weight daily. This view is consistent with the observation by Heinle (1966) that the related species *A. tonsa* can turnover its own biomass in as little as 2 days. However, it must be emphasized that the factor used in Table 8 for converting oxygen consumed into lipid metabolized is based on the value given by Hawk, Oser &

Summerson (1949) for the oxidation of palmitin: this of course is an over-simplification, as it is well known that zooplankton contain lipid material other than fatty acids, including hydrocarbons (see p. 407) and sterols (Gastaud, 1961; Fagerlund, 1962). Furthermore, there is evidence that the utilization of lipid by starving zooplankton is by no means common to all species. Thus, Orr (1934) found no significant change in the lipid levels of stage V *Calanus finmarchicus* during winter; Linford (1965) observed that these levels remained constant in three species of mysid starved for 4 days and in *C. helgolandicus* starved for 8 days; and Littlepage (1964) found that the lipid content of the carnivorous copepod *Euchaeta antarctica* remained fairly constant throughout most of the year, rising only during the period of egg-formation in summer. It would seem that, just as the total quantity of lipid in zooplankton varies from species to species (Fisher, 1962; Yamada, 1964), so too does the importance of lipid as a food reserve. There is also recent evidence (Conover & Corner, 1968) that for each of several species of zooplankton the use of lipid as a source of energy varies considerably with season: indeed, at certain times of year the high rate of nitrogen excretion implies that mainly protein is being catabolized.

Qualitative studies   (3) *Excretion of nitrogen*

It has generally been assumed that zooplankton are ammonotelic and investigations of their nitrogen excretion have usually involved measurements of ammonia liberated by these animals (Harris, 1959; Beers, 1964). The validity of this assumption, at least for *Calanus finmarchicus* and *C. helgolandicus*, was verified by some experiments of Corner, Cowey & Marshall (1965). Sea water which had contained *Calanus* was analysed for dissolved nitrogenous end-products by (i) nesslerization, (ii) steam distillation in a Kjeldahl apparatus, (iii) the ninhydrin method of Moore & Stein (1954). The values obtained by all three methods were sensibly the same. Because methods (i) and (ii) measured only ammonia while ninhydrin would estimate a variety of other compounds in addition to ammonia (e.g. amino acids, amines, amino sugars, etc.), it was concluded that ammonia was the main nitrogenous end-product. Later, Corner & Newell (1967), using a further variety of differential chemical methods, confirmed that *C. helgolandicus* is primarily ammonotelic, excreting only small and variable amounts of other nitrogenous substances, True, earlier investigations with larger aquatic invertebrates had indicated that considerable quantities of amino acids were excreted by these animals (for summaries of values see Nicol, 1960; Parry, 1960; Prosser & Brown, 1962), but the investigations were made about 35 years ago when reliable methods for estimating amino acids were not available. In a later study (Dresel & Moyle, 1950) it was found that a variety of marine littoral and estuarine amphipods and isopods were ammonotelic: small amounts of amino acid nitrogen were also detected among the excretory products, but the authors had considerable doubts as to whether the method used for measuring amino acid nitrogen gave meaningful results. Furthermore they found 'that a high percentage of casualties was often associated with exceptionally high values for amino acid nitrogen'. More recently, Needham (1957) has shown that 86 % of the nitrogen excreted by *Carcinides maenas* is ammonia and he could find no evidence for the excretion of amino acids.

Nevertheless, the view that zooplankton are primarily ammonotelic may need further consideration in the light of recent reports by Johannes & Webb (Johannes & Webb, 1965; Webb & Johannes, 1967) that large amounts of amino acids (from 2·2 to 47 mg. α-amino N/g. dry wt./day) are released by these animals. The authors claim that the rate of amino acid release is positively correlated with temperature according to an equation: release rate (mg. α-amino N/g. dry wt./day) = temperature (°C.) − 6·0.

However, in our view, these findings need to be treated with a certain caution. Thus, the experiments were carried out by placing zooplankton (0·8–11 g. dry wt./l.) in shallow containers of sea water at the collection temperature for periods ranging from 20 min. to 30 hr. The plankton were then collected on a membrane filter and the sea water subsequently analysed. It must be emphasized that this is an extraordinarily high concentration of zooplankton. Thus, it represents from 5000 to 70,000 *Calanus finmarchicus* (the animal with which we are most familiar) in each litre. It is difficult to imagine that such a paste of animals can be collected on a membrane filter without damage to the cells and concomitant release and filtration of free amino acids. In this respect it is worth noting that the dissolved free amino acids found by Webb and Johannes are predominantly glycine, taurine and alanine—that is, those which predominate in the free amino acid fraction of the cells (Cowey & Corner, 1963*b*). Webb and Johannes quote the zooplankton density of surface waters as 0·001–0·2 g. dry wt./m.³, which is equivalent to 0·001 − 0·2/1000 g. dry wt./l. It follows that in their experiments the density of zooplankton used was at least 4000 times greater than that occurring naturally.

The authors give no information on the quantity of ammonia released by the zooplankton they examined so that one cannot directly deduce whether or not the animals were primarily ammonotelic. Instead, they apply their formula to the data of Harris (1959) and deduce that the amount of α-amino N released is equivalent to 26 % of the ammonia N excreted by his zooplankton (at 14° C.). They conclude that on average planktonic invertebrates apparently excrete ammonia and dissolved free amino acids in the ratio of approximately 4:1. However, Beers (1964) found that *Sagitta hispida* excreted 12·7 mg. ammonia N/g. dry wt./day at 20° C., and if the formula of Webb and Johannes is applied in this instance one would deduce that 14 mg. α-amino N/g. dry wt./day would be simultaneously excreted by this animal; that is, *S. hispida* would not be primarily ammonotelic. Furthermore, if the formula is applied to data for *Calanus helgolandicus* it leads to the conclusion that nearly all the nitrogen excreted by this animal is in the form of amino acids, which is contrary to all previous findings (see Corner & Newell, 1967).

It should also be made clear that the total amounts of amino acid N excreted (Table 1; Webb & Johannes, 1967) exceed the values for α-amino N because the basic amino acids (lysine, histidine and arginine) contain more than one N atom per mole. Thus, sample 9 (99 % copepods) excreted 12 mg. α-amino N/g. dry wt./day of which 5·6 % (0·67 mg.) was α-amino N in arginine. But the total arginine N liberated would have been 2·68 mg. (of which 0·67 mg. was α-amino N) and therefore values for nitrogen excreted as amino acids quoted by Johannes and Webb are even higher than would appear at first sight.

Quantitative studies

Considering now the quantitative aspects of nitrogen excretion based on measurements of ammonia production by zooplankton, Harris (1959) found that mixed zooplankton in Long Island Sound excreted an average of 36·4 μg. ammonia N/mg. dry wt./day. The animals contained 89 μg. nitrogen/mg. dry wt. and the average daily excretion therefore corresponds to 41 % of this figure: if it is assumed that some 90 % of the body nitrogen is protein (Cowey & Corner, 1963b) then the daily excretion of nitrogen by the animals corresponds to 45 % of their protein content. A much lower value was observed in experiments with the chaetognath *Sagitta hispida* by Beers (1964), who obtained average excretion rates of ammonia nitrogen at 20° C. corresponding to about 14·5 % of the total body nitrogen of the animal each day. In a more detailed study of nitrogen excretion by *Calanus*, Corner *et al.* (1965) found that food availability, sea-water temperature and body weight were all factors influencing the rate of nitrogen excretion by *C. helgolandicus* and *C. finmarchicus*. Under conditions of food availability and sea temperature similar to those prevailing in Long Island Sound at the time of Harris's study, adult female *Calanus* excreted 10–16 μg. N/mg. dry body wt./day, a range of values well below that found by Harris. However, it was observed in experiments with animals at different stages of development that the rate of nitrogen excretion varied also with the size of the animals and that as body weight decreased by a factor of five so the rate of nitrogen excretion doubled. This value was a first approximation only, and in a later study (Corner, Cowey & Marshall, 1967), in which this aspect of the problem was examined with naupliar and copepodite stages of *C. helgolandicus*, the factor was found to be 4·1. Obviously, many more data with numerous species of different sizes are needed before this relationship can be said to be established for zooplankton in general, but the present value partially reconciles the values of Harris with those of Corner *et al.* (1965). Thus, the excretion rate for Harris's animals (in most instances consisting mainly of *Acartia clausi* of approximately 5 μg. dry body wt.) was 36·4 μg. N/mg. dry wt.—roughly twice the value of 20·9 found by Corner *et al.* (1965) for young *Calanus* of approximately 5 times the dry body wt. (i.e. 24·6 μg./animal).

As the rate of nitrogen excretion varies with temperature and food level, seasonal variations in excretion rate are to be expected. Conover & Corner (1968) have shown that such variations occur, *Calanus finmarchicus*, *C. hyperboreus*, *Metridia longa* and *Pareuchaeta norvegica* having a similar seasonal pattern of nitrogen excretion, being high in spring and decreasing gradually through summer and autumn to a winter minimum. This pattern is very similar to the seasonal variation in oxygen consumption.

The rapidity with which ingested nitrogen in the form of protein can be excreted by zooplankton is surprisingly high when compared with that of larger marine invertebrates. Thus, Dresel and Moyle (1950) obtained values for amphipods of about 1·8 μg. nitrogen/g. dry wt./day and Needham (1957) gives a value of 44 μg. nitrogen/g. body wt./day for fasting *Carcinides maenas*, a daily loss of the same order of magnitude as that of a fasting man. Attention was drawn earlier (see p. 405) to various factors favouring the efficient synthesis of protein by zooplankton. By contrast, the results of excretion studies—especially those made with small, metabolically active species—

118

imply that amino acids (presumably a mixture of those absorbed from the gut, together with those released during dynamic exchange in the tissue proteins) are rapidly deaminated. Accordingly, although protein synthesis may be efficient in these animals, protein retention is not, and the over-all process by which protein is laid down by growing zooplankton may, on balance, be rather wasteful. Virtually nothing is known of the metabolic pathways involved in nitrogen metabolism in marine invertebrates. It may be assumed that many enzymic steps and some translocation must occur before food protein is hydrolysed, absorbed and then either laid down as tissue protein in the animal or deaminated with the excretion of ammonia. There is no hint of any adaptation at tissue, cellular or enzyme level which might explain the very high metabolic activity of which zooplankton seem to be capable.

(4) O:N ratio

A measure of the atomic ratio oxygen consumed: nitrogen excreted provides useful data on the nature of the substrate oxidized by zooplankton, a low value implying that protein is mainly used and a high value characterizing the breakdown of fat or carbohydrate. The significance of the ratio in studies of zooplankton nutrition was first demonstrated by Harris (1959), who obtained an average value of 7·7 for the mixed zooplankton in Long Island Sound, a figure he regarded as anomalous, nitrogen being excreted back into the sea more rapidly than carbon was oxidized. Recently, Corner et al. (1965) have shown that Harris's value of 7·7 is not, in fact anomalous. On the contrary, good agreement is found between the amount of carbon utilized (assessed from the values for nitrogen excreted) and the quantity of carbon combusted (estimated from the data for oxygen consumption). For *Calanus helgolandicus* and *C. finmarchicus*, Corner et al. (1965) found O:N ratios varying from 9·8 to 15·6 (mean 13·5), values only slightly lower than the figure of 17 calculated by Redfield et al. (1963) for the oxidation of a typical sample of particulate organic matter in the sea. However, in a more recent study (Conover & Corner, 1968) the O:N ratios of several species of zooplankton have been found to vary considerably with season. Thus, for *C. finmarchicus* collected from the Gulf of Maine, the O:N ratio in April was less than 20 but with the coming of the phytoplankton bloom it rose to a value greater than 50 in May and then declined throughout summer and autumn to a value less than 20 in November. From supplementary studies made in the laboratory Conover and Corner found no evidence of temperature affecting the O:N ratio and concluded that the type of seasonal variation in the ratio observed with *Calanus* spp. probably reflected changes in the levels of available food, the animals adjusting the relative proportions of fat and protein metabolized in response to unfavourable food conditions. They point out that the value of 7·7 observed by Harris was low enough to imply that protein was chiefly used as an energy source by his animals. The period of Harris's study (April–June) was observed by Conover (1956) to be the season of most severe competition for food between *Acartia tonsa* and *A. clausi*. It therefore seems likely that the animals studied by Harris were catabolizing protein at a time when food was relatively scarce.

119

Studies on the feeding efficiency of zooplankton have been concerned mainly with two aspects: (a) assimilation efficiency (i.e. the percentage of captured food assimilated), and (b) gross growth efficiency (i.e. the percentage of captured food invested in growth). It should be emphasized that these efficiency values are calculated from data obtained in laboratory experiments: they need to be differentiated from 'ecological efficiency', a term used by Slobodkin (1960) to describe the steady state ratio of yield (of predators) to food ingested (as prey) by communities of animals in their natural habitat.

(1) Assimilation efficiency

Various methods have been used to estimate the percentage of captured food digested by zooplankton, including (1) feeding the animals with isotope-labelled algal diets and subsequently measuring the radioactivity in the animal body, eggs and faecal pellets (Marshall & Orr, 1955 b, 1961; Lasker, 1960); (2) chemical analyses of the particulate material captured and of the faecal pellets produced (Corner, 1961; Cowey & Corner, 1966); and (3) determining the ratio between the organic fraction of the food and that of the faecal pellets (Conover, 1966a; Haq, 1967; Corner et al. 1967). All these methods are open to criticism. Thus, the tracer-isotope technique requires that complete and immediate mixing of the isotope and carrier occurs inside the animal: if several 'pools' of material are involved, each with a different turnover rate, this requirement will not be met (Conover, 1961). Furthermore, both this technique and the chemical methods also require the quantitative recovery of faecal material, which is not always possible. Conover's 'ratio' method, on the other hand, has the advantage that quantitative recovery of faecal material is not necessary: however, it depends on the assumption that only the organic component of the food is significantly affected by the digestive process. Experimental evidence to support this assumption is lacking; if inorganic ions are freely distributed in the cytoplasm of the algal cells they may be readily assimilable by the zooplankton. In addition, both algal food and faecal pellets are washed with isotonic ammonium formate which is later volatilized at 70° C. before dry weights and ash contents are determined. Whether or not any change in the inorganic content of either algal cells or faecal pellets occurs (as a result of diffusion exchange) during this washing process does not seem to have been checked. Nevertheless in some instances where values obtained by the 'ratio' method have been compared directly with those obtained by other techniques (Conover, 1966a; Corner et al. 1967) reasonably good agreement has been found.

In Table 8 the values obtained for assimilation efficiency in laboratory experiments are summarized. These values cover a very wide range and there is no general agreement as yet on the efficiency of assimilation in different members of the zooplankton. In addition to the differences which are expected to result from the application of these different (and somewhat empirical) methods to different species there is also the possibility that important fractions of the food are assimilated at different rates and there is a case for attempting to examine the assimilation of such fractions (proteins, fats and sugars) separately (Cowey & Corner, 1966).

There appear to be two main views on the efficiency of assimilation by zooplankton: (1) that of Conover and of Marshall and Orr who find that ingested food is assimilated very efficiently (from 60 % to over 90 %) even when the animals (*Calanus*) are feeding in a rich algal culture and producing faecal pellets rapidly; and (2) that of Beklemishev who holds that when high concentrations of food are available zooplankton ingest more food than they need but assimilate relatively little (i.e. 'superfluous feeding' is said to occur).

Table 8. *Assimilation of phytoplankton by various zooplankton*

species	food	method	assimilation (%)	reference
Calanus finmarchicus	various diatoms and flagellates	tracer-isotope [^{32}P]	15–99	Marshall & Orr (1961)
	Skeletonema costatum	tracer-isotope [^{14}C]	60–78	Marshall & Orr (1955b)
Temora longicornis	*S. costatum*	tracer-isotope [^{32}P]	50–98	Berner (1962)
Euphausia pacifica	*Dunaniella primolecta*	tracer-isotope [^{14}C]	85–99	Lasker (1960)
Ostrea edulis (larvae)	*Isochrysis galbana*	tracer-isotope [^{32}P]	13–50	Walne (1965)
Calanus helgolandicus	natural particulate material	chemical analyses	74–91	Corner (1961)
Metridia lucens	*T. nordensköldii*	'ratio' method	50–84 ⎫	
	Ditylum spp.		35 ⎬	Haq (1967)
	Artemia nauplii		59 ⎭	
Calanus hyperboreus	*Exuviella* spp.	'ratio method'	39·0–85·6 ⎫	
		chemical analyses	54·6–84·6 ⎪	
Natural zooplankton	natural particulate material	'ratio' method	32·5–92·1 ⎬	Conover (1966a)
			⎪	
			⎭	
Calanus finmarchicus	*Skeletonema costatum*	'ratio' method	53·8–64·4 ⎫	Corner, Cowey & Marshall
	S. costatum	chemical analyses	57·5–67·5 ⎭	(1967)

The findings of Conover and of Marshall and Orr are based on laboratory experiments using either tracer methods or the 'ratio' method (e.g. Conover, 1964, 1966b; Marshall & Orr, 1955b, 1962). Thus Conover (1966b) investigated the assimilation of food by *Calanus hyperboreus* fed a series of food concentrations bracketing the level at which superfluous feeding would occur in the sea (390 μg. particulate carbon/l.: Beklemishev, 1962) and found no correlation between percentage of food assimilated and either food concentration or the total food ingested. Similarly he found no relationship between the amount of food present in the sea at any given time and the percentage digested by mixed zooplankton. He proposed that zooplankton respond to sudden outbursts of phytoplankton by increasing their fecundity and building up their food reserves which they use later when food is scarce. Relevant to this is the observation by McLaren (1963) that food storage may be assisted by the diurnal (and seasonal) vertical migration of zooplankton in the sea between warmer, surface waters and deeper, cooler waters. In McLaren's view animals at the surface feed at temperatures higher than those governing growth and development. The uptake of food is more efficient at these higher surface temperatures while the energy for growth processes is more efficiently used when the animals return to deeper, cooler waters. This process is thought to provide an energy 'bonus' which is invested in greater fecundity and storage of food.

On the other hand the views of Beklemishev are supported largely by field observations made by him and by others (e.g. Harvey, Cooper, Lebour & Russell, 1935; Lucas, 1936; Beklemishev, 1957, 1962) in which zooplankton were found during spring blooms to consume 30–50 % of their own biomass daily. As calculations (Riley, 1947; Harvey, 1950) indicated that zooplankton require only about 10 % of their own body weight per day to compensate for respiratory losses and to ensure normal growth it is deduced that at these times zooplankton do not efficiently utilize their food; that is, assimilation is low.

Additional support for Beklemishev has come from the field observations of Cushing & Vucetic (1963). From a 3-month study of a *Calanus* patch in the North Sea, during which animal and plant biomasses were periodically measured and algal productive rates and grazing mortality rates periodically estimated, they derived the quantity of algae apparently eaten by *Calanus*. During the spring outburst of phytoplankton the daily quantity of food apparently consumed by *Calanus* amounted to more than 3 times their body weight. Because the 'true metabolic need' probably amounts to about half the body weight per day (this estimate is taken from tables of Winberg (1956) which relate to fish but which Cushing and Vucetic claim are also applicable to Crustacea) superfluous feeding was said to be experimentally confirmed and quantitatively defined. The disparity between Cushing & Vucetic (1963) on the one hand and Riley (1947) and Harvey (1950) in the amount of food required to maintain a normal growth rate is in our view a commentary on how little is known about the physiology of zooplankton. Harvey (1950) makes the point: 'It is unknown whether these lowly animals need the whole range of amino acids required by mammals, and it is unknown whether phytoplankton protein is deficient in any one or more amino acids required by zooplankton. Thus it is possible that the animals need to digest and absorb more vegetable tissue than the sum of the quantities burnt in respiration and built into their own growing bodies.' This is still very much the case except that we should not restrict the statement to amino acids and proteins but would extend it to include a whole range of accessory food factors. Cushing and Vucetic claim to have shown 'that the effective reproductive capacity is limited to this period of superfluous feeding'. This may indicate that *Calanus* is able to obtain all necessary dietary constituents in sufficient quantity only when an excess of phytoplankton is available.

Superfluous feeding, if it does occur, may not be so wasteful in terms of the whole eco-system as might at first appear. Thus, Beklemishev (1962) has pointed out that the faecal pellets may have high nutritive value; and analysis of faecal pellets from grazing *Calanus* showed that they contained considerable amounts of protein composed of a broad spectrum of amino acids (Cowey & Corner, 1966).

It is considered by Conover (1964) that Cushing and Vucetic used methods for collecting zooplankton that may have missed numbers of very small animals which have high food needs and might have contributed substantially to grazing. In addition Cushing and Vucetic did not take into account the fact that the biomass of the zooplankton measured at any given time could have been affected by both their own reproductive rates and predation by fish.

This conflict of evidence and opinion on assimilation by zooplankton will remain until more is known about the feeding behaviour of zooplankton and until the physiology of assimilation is studied in more detail. While much is known about the composition of the algal food, little effort has been made to obtain similar data for faecal material. Analyses of faecal pellets obtained from animals ingesting algal cells at different rates would provide qualitative information on the nature of the digestive process. As reliable methods (e.g. gas–liquid chromatography, thin-layer chromatography, automated amino acid analysis) are now available such an approach is feasible and would provide a new and useful starting-point.

(2) Gross efficiency of growth

The percentage of captured food converted into new tissue by the animal can be calculated from the relationship

$$DQ = W_t - W_0 + E + M,$$

where D is the percentage of captured food digested, Q is the quantity captured (expressed either as dry weight, calories, or as units of a particular chemical substance, e.g. nitrogen), $W_t - W_0$ is the new tissue laid down by the animal in a known time (expressed in similar terms), E is the quantity expended in metabolism (calculated from oxygen uptake, carbon dioxide production or nitrogen excretion) and M is the amount lost as moults. A detailed account of the use of this relationship in estimating the gross growth efficiency of a marine zooplanktonic animal (*Calanus helgolandicus*) has been given by Corner et al. (1967) and the results are included in Table 9, which is based on recent data described by Conover (1968).

The combined data show a fairly wide scatter of values, for which there are several possible explanations. Thus, growth efficiency might be expected to vary with the

Table 9. *Gross growth efficiencies of various zooplankton*

species	stage	method of estimation	gross growth efficiency	reference
Calanus hyperboreus	IV	dry weight	13·0	Conover (1964)
	V	dry weight	14·6–36·4	
	IV	calories	18	
	V	calories	20–50	
C. finmarchicus	Sub-adult	nitrogen	34	Corner, Cowey & Marshall (1967)
	adult (egg-production)	nitrogen	14	
Artemia salina	young stages	dry weight	26–79	Reeve (1963)
	adult (production of young)	dry weight	15–25	
Euphausia pacifica	adult	[14]C	11–74	Lasker (1960)
Sagitta hispida	adult	dry weight	2–11	Reeve (1966)

123

concentration of available food. At excessively low levels the energy expended in feeding is not adequately replaced by the food captured; as the concentration of food increases so does efficiency until a peak is reached; efficiency then falls, because the animal is ingesting more food than it can assimilate. Such a peak of growth efficiency at a certain food level has been reported by Reeve (1963) for *Artemia*. Data seem to be lacking for other salt-water species, but Richman (1958) made a similar observation with *Daphnia*.

There is also evidence that growth efficiency varies with stage of development. Thus, Slobodkin (1960) quotes values obtained for *Daphnia* by Armstrong that show a higher growth efficiency during the early stage of development; and Corner *et al.* (1967) have shown that pre-adult growth efficiency for *Calanus finmarchicus* is 34 %, whereas the efficiency of egg production is only 14 %. Similar variations in growth efficiency with stage of development have been reported by Reeve (1963) for *Artemia*, high efficiencies being characteristic of periods of most rapid growth.

Other factors affecting growth efficiency, found in experiments with *Artemia*, are temperature and salinity (Reeve, 1963), efficiency increasing with temperature over the range 5–30° C. and rising to a peak at a salinity of 35 %.

The values obtained can also vary with species. For example, Corner *et al.* (1967) have compared their data for *Calanus* with those found by Lasker (1966) using *Euphausia pacifica*, and have noted that although by far the largest fraction of assimilated foodstuff is lost in metabolism by both species (61·4 % for *Calanus*; 72·3 % for *Euphausia*), they differ considerably in the amount of assimilated food they lose as moults (1 % for *Calanus*; 17 % for *Euphausia*). This could be partly due to the fact that the euphausiid continues to moult after it has become an adult. Possibly, in the transference of energy through the aquatic food chain the taxonomic composition of the zooplankton is a factor of some importance: but its full significance can only be assessed when data have been obtained with many other species.

VII. SUMMARY

1. Phytoplanktonic algae vary in their value as food for zooplankton and no single algal food can meet the full nutritional needs of zooplanktonic animals. Perhaps this is because optimal amounts of essential micronutrients are not all present in any one alga. The dietary requirements of planktonic Crustacea, as far as they are known, bear some resemblance to those of vertebrates.

2. The proteins of phytoplankton are similar in amino acid composition to those of zooplankton. This circumstance should favour efficient synthesis of protein by the animal, for assuming amino acids are all released in the gut and absorbed at approximately the same rate, they will be presented to the tissues in roughly the right relative amounts for protein formation.

3. Zooplankton are able to alter the characteristics of the fatty acids present in their diet by elongating the carbon chain length and by increasing the degree of unsaturation.

4. Measurements of phosphorus and nitrogen excretion indicate that zooplankton are metabolically very active. Some of the very high rates of phosphorus excretion are

questioned and it is suggested that some portion of the phosphorus compounds liberated by zooplankton have passed straight through the gut without being assimilated. It is unlikely that all forms of organically bound phosphorus are equally rapidly assimilated and turned over by zooplankton.

5. Estimates of the rate of ammonia excretion by zooplankton differ markedly. This may be a matter of size/surface area of the animals concerned—smaller animals excreting more rapidly than larger animals. It has been claimed that α-amino nitrogen is released in considerable quantities by zooplankton but the evidence is not yet compelling.

6. There is considerable disagreement on the efficiency of food assimilation and conversion by zooplankton. One view is that irrespective of the quantity of phytoplankton ingested assimilation is uniformly high. The opposing view holds that when rapid ingestion of phytoplankton occurs the percentage assimilated falls. More information on the feeding behaviour of zooplankton and on the physiology of their digestive processes is required before this controversy can be satisfactorily resolved.

VIII. REFERENCES

ACKMAN, R. G., JANGAARD, P.M., HOYLE, R. J. & BROCKERHOFF, H. (1964). Origin of marine fatty acids. I. Analyses of the fatty acids produced by the diatom *Skeletonema costatum*. *J. Fish. Res. Bd Can.* 21, 747–56.

ACKMAN, R. G., TOCHER, C. S. & McLACHLAN, J. (1968). *J. Fish. Res. Bd Can.* (In the Press.)

BAALSRUD, K. (1955). Utilization of plankton. *Norsk Hvalfangsttid.* no. 3, pp. 125–33.

BARNES, H. & EVENS, R. (1967). Studies in the biochemistry of cirripede eggs. III. Changes in the amino-acid composition during development of *Balanus balanoides* and *B. balanus*. *J. mar. biol. Ass. U.K.* 47, 171–80.

BEERS, J. R. (1964). Ammonia and inorganic phosphorus excretion by the planktonic chaetognath, *Sagitta hispida* Conant. *J. Cons. perm. int. Explor. Mer* 29, 123–9.

BEERSTECHER, E. (1964). The biochemical basis of chemical needs. In *Comparative Biochemistry*. Vol. VI. *Cells and Organisms*, pp. 119–220. Ed. M. Florkin and H. S. Mason. New York and London: Academic Press.

BEKLEMISHEV, C. W. (1957). Superfluous feeding of zooplankton and the problem of sources of food for bottom animals. *Trud. Vsesoyuz. gidrobiol. Obshch.* 8, 354–8.

BEKLEMISHEV, C. W. (1962). Superfluous feeding of marine herbivorous zooplankton. *Rapp. P.-v. Réun. Cons. perm. int. Explor. Mer* 153, 108–13.

BERNARD, M. (1961). Adaptation of some Mediterranean pelagic copepods to survival media in an aquarium. *Rapp. Comm. int. Mer. Medit.* 16, 165–76.

BERNER, Å. (1962). Feeding and respiration in the copepod *Temora longicornis* (Müller). *J. mar. biol. Ass. U.K.* 42, 625–40.

BLUMER, M., MULLIN, M. M. & THOMAS, D. W. (1963a). Pristane in zooplankton. *Science, N.Y.* 140, 974.

BLUMER, M., MULLIN, M. M. & THOMAS, D. W. (1963b). Pristane in the marine environment. *Helg. wiss. Meeres.* 10, 187–201.

CANDY, D. J. & KILBY, B. A. (1962). Studies on chitin synthesis in the desert locust. *J. exp. Biol.* 39, 129–40.

CAREY, F. G. (1965). Chitin synthesis *in vitro* by crustacean enzymes. *Comp. Biochem. Physiol.* 16, 155–8.

CHAU, Y. K., CHUECAS, L. & RILEY, J. P. (1967). The component combined amino acids of some marine phytoplanktonic species. *J. mar. biol. Ass. U.K.* 47, 543–54.

CHRISTOMANOS, A. A., DIMITRIADIS, A. & GARDIKI, V. (1962). Contribution to plankton chemistry: I. *Chim. Chron.* 27 A, 23–6.

COLE, H. A. (1937). Experiments in the breeding of oysters *Ostrea edulis* in tanks, with special reference to the food of the larva and spat. *Fishery Invest., Lond.*, ser. 2. 15, no. 4.

CONOVER, R. J. (1956). Oceanography of Long Island Sound 1952–54. VI. Biology of *Acartia clausi* and *A. tonsa*. *Bull. Bingham oceanogr. Coll.* 15, 156–233.

CONOVER, R. J. (1960). The feeding behaviour and respiration of some planktonic copepods. *Biol. Bull. mar. biol. Lab.*, *Woods Hole* **119**, 399–415.

CONOVER, R. J. (1961). The turnover of phosphorus by *Calanus finmarchicus*. Addendum to paper by S. M. Marshall and A. P. Orr. On the biology of *Calanus finmarchicus*. XII. The phosphorus cycle: excretion, egg production, autolysis. *J. mar. biol. Ass. U.K.* **41**, 484–8.

CONOVER, R. J. (1962). Metabolism and growth in *Calanus hyperboreus* in relation to its life cycle. *Rapp. P.-v. Réun. Cons. perm. int. Explor. Mer* **153**, 190–7.

CONOVER, R. J. (1964). Food relations and nutrition of zooplankton. *Symp. exp. mar. Ecol.*, *in Narragansett mar. Lab. occ. Publ.* no. 2, pp. 81–9.

CONOVER, R. J. (1965). Notes on the moulting cycle, development of sexual characters and sex ratio in *Calanus hyperboreus*. *Crustaceana* **8**, 308–20.

CONOVER, R. J. (1966a). Assimilation of organic matter by zooplankton. *Limnol. Oceanogr.* **11**, 338–45.

CONOVER, R. J. (1966b). Factors affecting the assimilation of organic matter by zooplankton and the question of superfluous feeding. *Limnol. Oceanogr.* **11**, 346–54.

CONOVER, R. J. (1968). Zooplankton—life in a nutritionally dilute environment. *Am. Zool.* **8**, 107–18.

CONOVER, R. J. & CORNER, E. D. S. (1968). Respiration and nitrogen excretion by some marine zooplankton in relation to their life cycles. *J. mar. biol. Ass. U.K.* **48**, 49–75.

CORNER, E. D. S. (1961). On the nutrition and metabolism of zooplankton. I. Preliminary observations on the feeding of the marine copepod, *Calanus helgolandicus* (Claus). *J. mar. biol. Ass. U.K.* **41**, 5–16.

CORNER, E. D. S. & COWEY, C. B. (1964). Some nitrogenous constituents of the plankton. *Oceanogr. mar. Biol. ann. Rev.* **2**, 147–67. Ed. H. Barnes. London: Allen and Unwin.

CORNER, E. D. S., COWEY, C. B. & MARSHALL, S. M. (1965). On the nutrition and metabolism of zooplankton. III. Nitrogen excretion by *Calanus*. *J. mar. biol. Ass. U.K.* **45**, 429–42.

CORNER, E. D. S., COWEY, C. B. & MARSHALL, S. M. (1967). On the nutrition and metabolism of zooplankton. V. Feeding efficiency of *Calanus finmarchicus*. *J. mar. biol. Ass. U.K.* **47**, 252–70.

CORNER, E. D. S. & NEWELL, B. S. (1967). On the nutrition and metabolism of zooplankton. IV. The forms of nitrogen excreted by *Calanus*. *J. mar. biol. Ass. U.K.* **47**, 113–20.

COWEY, C. B. (1956). A preliminary investigation of the variation of vitamin B_{12} in oceanic and coastal waters. *J. mar. biol. Ass. U.K.* **35**, 609–20.

COWEY, C. B. & CORNER, E. D. S. (1962). The amino acid composition of *Calanus helgolandicus* (Claus) in relation to that of its food. *Rapp. P.-v. Réun. Cons. perm. int. Explor. Mer* **153**, 124–8.

COWEY, C. B. & CORNER, E. D. S. (1963a). On the nutrition and metabolism of zooplankton. II. The relationship between the marine copepod *Calanus helgolandicus* and particulate material in Plymouth sea water in terms of amino acid composition. *J. mar. biol. Ass. U.K.* **43**, 495–511.

COWEY, C. B. & CORNER, E. D. S. (1963b). Amino acids and some other nitrogenous constituents in *Calanus finmarchicus*. *J. mar. biol. Ass. U.K.* **43**, 485–93.

COWEY, C. B. & CORNER, E. D. S. (1966). The amino acid composition of certain unicellular algae and of the faecal pellets produced by *Calanus finmarchicus* when feeding on them. In *Some contemporary studies in marine science*, pp. 225–31. Ed. H. Barnes. London: Allen and Unwin.

CUSHING, D. H. & VUCETIC, T. (1963). Studies on a *Calanus* patch. III. The quantity of food eaten by *Calanus finmarchicus*. *J. mar. biol. Ass. U.K.* **43**, 349–71.

DAVIS, H. C. & GUILLARD, R. R. (1958). Relative value of ten genera of micro-organisms as foods for oyster and clam larvae. *Fishery Bull. Fish Wildl. Serv. U.S.* **58**, 293–304.

DRESEL, E. I. B. & MOYLE, V. (1950). Nitrogenous excretion of amphipods and isopods. *J. exp. Biol.* **27**, 210–25.

ELLENBOGEN, L. (1963). Vitamin B_{12} and Intrinsic Factor. In *Newer methods of nutritional biochemistry*, vol. I, pp. 235–87. Ed. A. A. Albanese. New York and London: Academic Press.

FAGERLUND, U. H. M. (1962). Marine sterols. X. The sterol content of the plankton *Euphausia pacifica*. *Can. J. Biochem. Physiol.* **40**, 1839–40.

FARKAS, T. & HERODEK, S. (1964). The effect of environmental temperature on the fatty acid composition of crustacean plankton. *J. Lipid Res.* **3**, 369–73.

FISHER, L. R. (1962). The total lipid material in some species of marine zooplankton. *Rapp. P.-v. Réun. Cons. perm. int. Explor. Mer* **153**, 129–36.

FORD, J. E. (1960). Microbiological assay of protein quality. *Br. J. Nutr.* **14**, 485–98.

GASTAUD, J. M. (1961). Contributions to the biochemistry of the plankton. Study on the unsaponifiable lipids and the sterol fraction. *Rapp. Comm. int. Mer Medit.* **16**, 201–4.

GIBOR, A. (1956). Some ecological relationships between phyto- and zooplankton. *Biol. Bull. mar. biol. Lab.*, *Woods Hole* **111**, 230–4.

HALVER, J. E. (1957). Nutrition of Salmonoid fishes. IV. An amino acid test diet for Chinook Salmon. *J. Nutr.* **62**, 245–54.

HALVER, J. E., DELONG, D. C. & MERTZ, E. T. (1957). Nutrition of Salmonoid fishes. V. Classification of essential amino acids for Chinook Salmon. *J. Nutr.* **63**, 95–105.

HAQ, S. M. (1967). Nutritional physiology of *Metridia lucens* and *Metridia longa* from the Gulf of Maine. *Limnol. Oceanogr.* **12**, 40–51.

HARRIS, E. (1959). The nitrogen cycle in Long Island Sound. *Bull. Bingham. oceanogr. Coll.* **17**, 31–65.

HARRIS, E. & RILEY, G. A. (1956). Oceanography of Long Island Sound, 1952–1954. VIII. Chemical composition of the plankton. *Bull. Bingham oceanogr. Coll.* **15**, 315–23.

HARVEY, H. W. (1950). On the production of living matter in the sea off Plymouth. *J. mar. biol. Ass. U.K.* **29**, 97–137.

HARVEY, H. W., COOPER, L. H. N., LEBOUR, M. V. & RUSSELL, F. S. (1935). Plankton production and its control. *J. mar. biol. Ass. U.K.* **20**, 407–42.

HAWK, P, B., OSER, B. L. & SUMMERSON, W. H. (1949). *Practical physiological chemistry*, 12th ed. Philadelphia and Toronto: The Blakiston Company.

HEINLE, D. R. (1966). Production of a Calanoid copepod *Acartia tonsa* in the Patukent River estuary. *Chesapeake Sci.* **7**, 59–74.

HUTNER, S. H. & PROVASOLI, L. (1965). Comparative physiology: Nutrition. *A. Rev. Physiol.* **27**, 19–50.

HUTCHISON, W. C. & MUNRO, H. N. (1961). The determination of nucleic acids in biological materials. *The Analyst* **86**, 768–813.

JOHANNES, R. E. (1964*a*). Phosphorus excretion and body size in marine animals: microzooplankton and nutrient regeneration. *Science, N.Y.* **146**, 923–4.

JOHANNES, R. E. (1964*b*). Uptake and release of phosphorus by a benthic marine amphipod. *Limnol. Oceanogr.* **9**, 235–42.

JOHANNES, R. E. & WEBB, K. L. (1965). Release of dissolved amino acids by marine zooplankton. *Science, N.Y.* **150**, 76–7.

KASTING, R. & MCGINNIS, A. J. (1958). Use of glucose labelled with ^{14}C to determine the amino acids essential for an insect. *Nature, Lond.* **182**, 1380–81.

KAYAMA, M., TSUCHIYA, Y. & MEAD, J. F. (1963). A model experiment of aquatic food chain with special significance in fatty acid conversion. *Bull. Jap. Soc. sci. Fish.* **29**, 452–8.

KETCHUM, B. H. (1962). Regeneration of nutrients by zooplankton. *Rapp. P.-v. Réun. Cons. perm. int. Explor. Mer* **153**, 142–7.

LASKER, R. (1960). Utilization of organic carbon by a marine crustacean: analysis with [14]C. *Science, N.Y.* **131**, 1098–1100.

LASKER, R. (1966). Feeding, growth, respiration and carbon utilization of a Euphausiid crustacean. *J. Fish. Res. Bd Can.* **23**, 1291–1317.

LASKER, R. & THEILACKER, G. H. (1965). Maintenance of euphausiid shrimps in the laboratory. *Limnol. Oceanogr.* **10**, 287–8.

LEWIS, A. G. (1967). An enrichment solution for culturing the early developmental stages of the planktonic marine copepod *Euchaeta japonica* (Marakawa). *Limnol. Oceanogr.* **12**, 147–8.

LINFORD, E. (1965). Biochemical studies on marine zooplankton. II. Variations in the lipid content of some *Mysidacea*. *J. Cons. perm. int. Explor. Mer* **30**, 16–27.

LITTLEPAGE, J. L. (1964). Seasonal variation in lipid content of two Antarctic marine Crustacea. *Biologie Antarctique*, pp. 463–70. Ed. R. Carrick. Paris: Hermann.

LOOSANOFF, V. L., DAVIS, H. C. & CHANLEY, P. E. (1955). Food requirements of some bivalve larvae. *Proc. natn. Shellfish. Ass.* **45**, 66–83.

LOVERN, J. A. (1935). Fat metabolism in fishes. VI. The fats of some plankton Crustacea. *Biochem. J.* **29**, 847–9.

LOVERN, J. A. (1936). Fat metabolism in fishes. IX. The fats of some aquatic plants. *Biochem. J.* **30**, 387–90.

LUCAS, C. E. (1936). On certain inter-relations between phytoplankton and zooplankton under experimental conditions. *J. Cons. int. Explor. Mer* **12**, 343–62.

MAGENDIE, F. (1816). Sur les propriétés nutritives des substances qui ne contiennent pas d'azote. *Ann. Chim. Phys.* **3**, 66–77.

MANGOLD, H. K. & SCHLENK, H. (1957). Preparation and isolation of fatty acids randomly labelled with C[14]. *J. biol. Chem.* **229**, 731–41.

MARSHALL, S. M. & ORR, A. P. (1952). On the biology of *Calanus finmarchicus*. VII. Factors affecting egg production. *J. mar. biol. Ass. U.K.* **30**, 527–48.

MARSHALL, S. M. & ORR, A. P. (1955*a*). On the biology of *Calanus finmarchicus*. VIII. Food uptake, assimilation and excretion in adult and stage V *Calanus*. *J. mar. biol. Ass. U.K.* **34**, 495–529.

MARSHALL, S. M. & ORR, A. P. (1955*b*). Experimental feeding of the copepod *Calanus finmarchicus* (Gunnerus) on phytoplankton cultures labelled with radio-active carbon (C[14]). Suppl. *Deep-Sea Res.* **3**, 110–14.

MARSHALL, S. M. & ORR, A. P. (1958). On the biology of *Calanus finmarchicus*. X. Seasonal changes in oxygen consumption. *J. mar. biol. Ass. U.K.* **37**, 459–72.

MARSHALL, S. M. & ORR, A. P. (1961). On the biology of *Calanus finmarchicus*. XII. The phosphorus cycle; excretion, egg production, autolysis. *J. mar. biol. Ass. U.K.* **41**, 463–88.

MARSHALL, S. M. & ORR, A. P. (1962). Food and feeding in copepods. *Rapp. P.-v. Réun. Cons. perm. int. Explor. Mer* **153**, 92–8.

MCLACHLAN, J. & CRAIGIE, J. S. (1966). Chitan fibres in *Cyclotella cryptica* and growth of *C. cryptica* and *Thalassiosira fluviatilis*. In *Some contemporary studies in marine science*, pp. 511–17. Ed. H. Barnes. London: Allen and Unwin.

MCLACHLAN, J., MCINNES, A. G. & FALK, M. (1965). Studies on the chitan (chitin, poly-*N*-acetylglucosamine) fibres of the diatom *Thalassiosira fluviatilis* Hustedt. I. Production and isolation of chitin fibers. *Can. J. Bot.* **43**, 707–13.

MCLAREN, I. A. (1963). Effects of temperature on growth of zooplankton and the adaptive value of vertical migration. *J. Fish. Res. Bd Can.* **20**, 685–727.

MOORE, S. & STEIN, W. H. (1954). A modified ninhydrin method for the photometric determinations of amino acids and related compounds. *J. biol. Chem.* **211**, 907–13.

MORIMURA, Y. (1959). Synchronous culture of *Chlorella*. 2. Changes in content of various vitamins during course of the algal life cycle. *Pl. Cell. Physiol.*, *Tokyo* **1**, 63–9.

MOYSE, J. (1963). A comparison of the value of various flagellates and diatoms as food for barnacle larvae. *J. Cons. int. Explor. Mer* **28**, 175–87.

MULLIN, M. M. & BROOKS, E. R. (1967). Laboratory culture, growth rate and feeding behaviour of a planktonic marine copepod. *Limnol. Oceanogr.* **12**, 657–66.

NEEDHAM, A. E. (1957). Factors affecting nitrogen excretion in *Carcinides maenas* (Pennant). *Physiol. comp. Oecol.* **4**, 209–39.

NEUNES, H. W. & PONGOLINI, G-F. (1965). Breeding a pelagic copepod *Euterpina acutifrons* (Dana), in the Laboratory. *Nature, Lond.* **208**, 571–3.

NICOL, J. A. C. (1960). *The biology of marine animals.* London: Pitman.

ODUM, E. P. (1961). Excretion rate of radioisotopes as indices of metabolic rates in nature: biological half life of Zn^{65} in relation to temperature, food consumption, growth and reproduction. *Biol. Bull. mar. biol. Lab.*, *Woods Hole* **121**, 371–2.

OGINO, C. (1963). Studies on the chemical composition of some natural foods of aquatic organisms. *Bull. Jap. Soc. sci. Fish.* **29**, 459–62.

ORR, A. P. (1934). On the biology of *Calanus finmarchicus*. IV. Seasonal change in the weight and chemical composition in Loch Fyne. *J. mar. biol. Ass. U.K.* **19**, 613–32.

OSER, B. L. (1959). An integrated essential amino acid index for predicting the biological value of proteins. In *Protein and amino acid nutrition*, p. 281. Ed. A. A. Albanese. New York and London: Academic Press.

PARRY, G. (1960). Excretion. In *The physiology of Crustacea*, vol. I, pp. 341–66. Ed. T. H. Waterman. New York: Academic Press.

PARSONS, T. R., STEPHENS, K. & STRICKLAND, J. D. H. (1961). On the chemical composition of eleven species of Marine Phytoplankters. *J. Fish. Res. Bd Can.* **18**, 1001–16.

PETIPA, T. S. (1966). Oxygen consumption and food requirement in copepods *Acartia clausi* Giesbr. and *A. latisetosa* Kritoz. *Zool. Zhurnal.* **45**, 363–70.

POMEROY, R., MATHEWS, H. M. & MIN, H. S. (1963). Excretion of phosphate and soluble organic phosphorus compounds by zooplankton. *Limnol. Oceanogr.* **8**, 50–5.

PROSSER, C. L. & BROWN, F. A. Jr. (1962). *Comparative animal physiology*, 2nd ed. Philadelphia: W. B. Saunders Co.

PROVASOLI, L. & D'AGOSTINO, A. (1962). Vitamin requirements of *Artemia salina* in aseptic culture. *Am. Zool.* **2**, 12.

PROVASOLI, L. & SHIRAISHI, K. (1959). Axenic cultivation of the brine shrimp *Artemia salina*. *Biol. Bull. mar. biol. Lab.*, *Woods Hole* **117**, 347–55.

PROVASOLI, L., SHIRAISHI, K. & LANCE, J. R. (1959). Nutritional idiosyncrasies of *Artemia* and *Tigriopus* in monoxenic culture. *Ann. N.Y. Acad Sci.* **77**, 250–61.

RAYMONT, J. E. G., AUSTIN, J. & LINFORD, E. (1964). Biochemical studies on marine zooplankton. I. The biochemical composition of *Neomysis integer*. *J. Cons. perm. int. Explor. Mer* **28**, 354–63.

RAYMONT, J. E. G., AUSTIN, J. & LINFORD, E. (1966a). Biochemical studies on marine zooplankton. III. Seasonal variation in the biochemical composition of *Neomysis integer*. In *Some contemporary studies in marine science*, pp. 597–605. Ed. H. Barnes. London: Allen and Unwin.

RAYMONT, J. E. G., AUSTIN, J. & LINFORD, E. (1966b). The biochemical composition of certain oceanic zooplanktonic decapods. *Deep-sea Res.* **14**, 113–15.

RAYMONT, J. E. G. & CONOVER, R. J. (1961). Further investigations on the carbohydrate content of marine zooplankton. *Limnol. Oceanogr.* **6**, 154–64.
RAYMONT, J. E. G. & GAULD, D. T. (1951). The respiration of some planktonic copepods. *J. mar. biol. Ass. U.K.* **29**, 681–93.
RAYMONT, J. E. G. & LINFORD, E. (1966). A note on the biochemical composition of some Mediterranean zooplankton. *Int. Rev. ges. Hydrobiol.* **51**, 485–8.
REDFIELD, A. C., KETCHUM, B. H. & RICHARDS, F. A. (1963). The influence of organisms on the composition of sea water. In *The Sea: Ideas and observations on progress in the study of the seas*. vol. II, pp. 26–77. Ed. M. N. Hill. New York: Interscience.
REEVE, M. R. (1963). Growth efficiency in *Artemia* under laboratory conditions. *Biol. Bull. mar. biol. Lab., Woods Hole* **125**, 133–45.
REEVE, M. R. (1966). Observations on the biology of a chaetognath. In *Some contemporary studies in marine science*, pp. 613–30. Ed. H. Barnes. London: Allen and Unwin.
RICHMAN, S. (1958). The transformation of energy in *Daphnia pulex*. *Ecol. Monogr.* **28**, 273–91.
RILEY, G. A. (1947). A theoretical analysis of the zooplankton population of Georges Bank. *J. mar. Res.* **6**, 104–13.
SATOMI, M. & POMEROY, L. R. (1965). Respiration and phosphorus excretion in some marine populations. *Ecology* **46**, 877–81.
SCHWIMMER, M. & SCHWIMMER, D. (1955). *The role of algae and plankton in medicine*. New York and London: Grune and Stratton.
SHIRAISHI, K. & PROVASOLI, L. (1959). Growth factors as supplement to inadequate algal foods for *Tigriopus japonicus*. *Tohoku J. agr. Res.* **10**, 89–96.
SLOBODKIN, L. B. (1960). Ecological energy relationships at the population level. *Am. Nat.* **94**, 213–36.
TEAL, J. M. & HALCROW, K. (1962). A technique for measurement of respiration of single copepods at sea. *J. Cons. perm. int. Explor. Mer* **27**, 125–8.
URRY, D. L. (1965). Observations on the relationship between the food and survival of *Pseudocalanus elongatus* in the laboratory. *J. mar. biol. Ass. U.K.* **45**, 49–58.
VACELET, E. (1961). La consommation d'oxygene du copepode *Tigriopus fulvus* (Fischer). *Recl. Trav. Stn mar. Endoume* **36**, 111–13.
WALNE, P. R. (1956). Experimental rearing of the larvae of *Ostrea edulis* L. in the laboratory. *Fish. Invest., Lond.*, ser. 2, **20** (9), 1–23.
WALNE, P. R. (1965). Observations on the influence of food supply and temperature on the feeding and growth of the larvae of *Ostrea edulis* L. *Fish Invest., Lond.*, ser. II, **24**, no. 1.
WEBB, K. L. & JOHANNES, R. E. (1967). Studies of the release of dissolved free amino acids by marine zooplankton. *Limnol. Oceanogr.* **12**, 376–82.
WINBERG, G. G. (1956). Rate of metabolism and food requirements of fishes. *Nauch. Trud. Belorusskovo Gos. Univ. meni Minsk*, 1956.
YAMADA, M. (1964). The lipid of plankton. *Bull. Jap. Soc. sci. Fish.* **30**, 673–81

IX. ADDENDUM

In a recent investigation of the organic material retained by aquatic invertebrates, Johannes & Satomi (1967) have criticized earlier studies on food assimilation. They point out that not all the organic matter present in faeces may come from the food: some may be secreted by the animal (e.g. mucus, chitinous secretions). They also suggest that not all the unassimilated food is released as faeces: some may be released in solution. Using the benthic, estuarine decapod *Palaemontes pugio* fed on *Nitzschia closterium*, Johannes and Satomi have determined the percentage of ingested food retained for growth and respiration by subtracting the total amount released, both as particulate and dissolved organic carbon, from the input of organic carbon as food. They obtained figures of 47 and 49% in their two experiments, and also found that more organic carbon was released in the dissolved state than as faeces. They maintain that, compared with these new data, earlier values for the assimilation of carbon have less ecological and physiological meaning. It is not yet known whether the high rate of release of dissolved organic carbon observed with *Palaemontes* is typical of marine invertebrates, including zooplankton.

Martin (1968) has described an inverse relationship between the metabolic rate of zooplankton and the level of available food. The animals, mainly *Acartia tonsa* and *A. clausi* collected from Narragansett Bay, showed minimal rates of ammonia and phosphate excretion when phytoplankton was plentiful (April–June) and maximal rates when phytoplankton was scarce (August–November). Martin concludes that when food is abundant carbohydrate is mainly used as an energy source, protein and lipid being stored: shortage of food, by contrast, leads to the use of both protein and lipid reserves as sources of energy, with concomitant increases in ammonia and phosphate excretion. However, no direct evidence is presented linking phosphate excretion with lipid catabolism, nor was any measurement made of carbohydrate present in the phytoplankton at different seasons. Moreover, food levels were estimated only in terms of cell counts. The quantities of foodstuffs available during the autumn in the form of particulate materials other than phytoplankton, such as detritus, were not included.

Kittredge, Horiguchi & Williams (1968) have demonstrated the biosynthesis in marine phytoplankton of several compounds containing the C—P bond, namely 2-aminoethylphosphonic acid (together with its mono-, di- and tri-N-methyl derivatives) and 2-amino-3-phosphopropionic acid. The authors have also shown that relatively higher amounts of 2-aminoethylphosphonic acid are present in the planktonic amphipod *Amonyx nugax*, as well as in a net sample of phytoplankton and microzooplankton. They draw attention to the possibility that zooplankton may selectively concentrate the aminophosphonic acids obtained in their food, presumably incorporating them into tissue phospholipids where they would take the place of the nitrogenous bases commonly found in these compounds. The C—P bond is very stable chemically, and there is no evidence that it can be split enzymically. Possibly, therefore, substances containing this bond are metabolically inert and could be used, in the same way as the hydrocarbon pristane (see p. 407), as 'indicators' of the quantity of phytoplankton assimilated by the animals.

The rearing from egg to adult of the calanoid copepods *Temora longicornis* and *Pseudocalanus minutus* has recently been described by Corkett (1967). The sea water used was taken from the same area of the sea as the animals, which were fed on a high concentration of *Isochrysis galbana*. Care was taken to see that the animals were handled with minimal disturbance, one of the main factors in the successful rearing of *A. tonsa* and *A. clausi* through multiple generations by Zillioux & Wilson (1966).

REFERENCES

CORKETT, C. J. (1967). Technique for rearing marine calanoid copepods in laboratory conditions *Nature, Lond.* **216**, 58–9.
JOHANNES, R. E. & SATOMI, M. (1967). Measuring organic matter retained by aquatic invertebrates. *J. Fish. Res. Bd Can.* **24**, 2467–71.
KITTREDGE, J. S., HORIGUCHI, M. & WILLIAMS, P. M. (1968). Aminophosphonic acids: biosynthesis by marine phytoplankton. *Science, N.Y.* (in the press).
MARTIN, J. H. (1968). Phytoplankton–zooplankton relationships in Narragansett Bay. III. Seasonal changes in zooplankton excretion rates in relation to phytoplankton abundance. *Limnol. Oceanogr.* **13**, 63–71.
ZILLIOUX, E. J. & WILSON, D. F. (1966). Culture of a planktonic calanoid copepod through multiple generations. *Science, N.Y.* **151**, 996–7.

131

AMINOPHOSPHONIC ACIDS: BIOSYNTHESIS BY MARINE PHYTOPLANKTON

J. S. KITTREDGE, M. HORIGUCHI and P. M. WILLIAMS

INTRODUCTION

THE PHOSPHONIC acid analog of taurine, 2-aminoethylphosphonic acid (2-AEP), occurs free and as a major constituent of the phospholipids of ciliates, coelenterates and molluscs (Horiguchi, 1967; Quin, 1967). Since the isolation of 2-AEP (Horiguchi & Kandatsu, 1959; Kittredge et al., 1962), four other aminophosphonic acids structurally related to 2-AEP have been discovered in coelenterates (Kittredge & Hughes, 1964; Kittredge et al., 1967). These were phosphonoalanine (2-amino-3-phosphonopropionic acid) and the three N-methyl derivatives of 2-AEP. Recently we isolated 2-AEP from the hydrolysates of a planktonic amphipod, Amonyx nugax, and from a net sample composed of phytoplankton and microzooplankton. The 2-AEP phosphorus comprised 2·9 and 3·2 per cent, respectively, of their total phosphorus. This suggested a planktonic origin and a possible concentration of the aminophosphonic acids in filter-feeding marine invertebrates. We report on six species of marine phytoplankters which have been examined for their ability to synthesize aminophosphonic acids.

MATERIALS AND METHODS

Three species of dinoflagellates, *Amphidinium carteri*, *Exuviella cassubica* and *Peridinium trochoidum*, and two species of coccolithophorids, *Coccolithus huxleyi* and *Syracosphaera elongata*, were cultured axenically in 1500 ml of modified von Stosch media (von Stosch & Drebs, 1964). The cultures were grown at 17°C under cool fluorescent light. After 15 days' growth 2 mc of ^{32}P-orthophosphate was added aseptically to each culture. The organisms were allowed to incorporate the label for 10 hr in one experiment and 88 hr in another (approximately one-half and four divisions).

The cells were harvested by centrifugation at 1500 g, rinsed once with fresh media and recentrifuged. The pellets were taken up in 20 ml of 6 N HCl and hydrolyzed in sealed tubes at 110°C for 48 hr to release all of the ester phosphate and aminophosphonic acids. After removal of the HCl, the hydrolysates were taken up in water, decolorized with Norit A, concentrated, adjusted to pH 2 and passed through a 1 × 12 cm column of Dowex-50H⁺ resin made up in 1 N formic acid. The columns were washed with 20 ml of 1 N formic and then eluted with water to neutrality to obtain the phosphonoalanine. Elution of the columns with 3 N NH₄OH released the 2-AEP and the N-methyl derivatives along with the amino acids. This eluate was concentrated and passed through a 1 × 4 cm column of Dowex-1-acetate with 0·1 N acetic acid to remove the residual orthophosphate and aliquots were taken for counting. Each fraction was concentrated and aliquots were subjected to paper electrophoresis in 1 N formic acid at 17 V/cm for 3·5 hr. The strips were scanned with a 4π counter and the peaks of activity were located and eluted. The identification of phosphonoalanine was confirmed by paper chromatography in a butanol–acetic acid–water (12 : 3 : 5) system. The "2-AEP" peaks exhibited shoulders indicating considerable amounts of N-methyl derivatives. Electrophoresis in the same system for 21 hr separated the N-trimethyl-2-AEP and N-dimethyl-2-AEP. The incompletely resolved N-methyl-2-AEP and 2-AEP were eluted and separated by paper chromatography in a phenol–water (4 : 1) system. The identity of the other two N-methyl-2-AEP's was confirmed with this system. Aliquots of each isolated aminophosphonic acid were subjected to rehydrolysis and reisolated to confirm the absence of residual phosphate esters.

RESULTS

Aminophosphonic acids were detected in all cultures. Three species exhibited high relative amounts of phosphonoalanine. Both species of coccolithophorids contained considerable N-dimethyl-2-AEP and only detectable amounts of the N-methyl-2-AEP, which is the reverse of the observations with coelenterates and ciliates (and now dinoflagellates). The relative amounts of the aminophosphonic acids isolated from these organisms are shown in Table 1.

TABLE 1—PERCENTAGE CONCENTRATIONS OF AMINOPHOSPHONIC ACIDS IN MARINE PHYTOPLANKTON

| | Aminophosphonic acids (%) | | | | |
	Pal	MM	DM	TM	2-AEP
Dinoflagellates					
Amphidinium carteri	21·5	18	>1	28	32·5
Peridinium trochoidem	—	—	—	5	95
Exuviella cassubica	3	—	—	—	97
Coccolithophorids					
Coccolithus huxleyi	24	<0·5	4	3	69
Syracosphaera elongata	24	<0·5	4	20	52

Abbreviations: 2-AEP, 2-aminoethylphosphonic acid; MM, DM and TM, the mono-, di- and tri-N-methyl derivatives of 2-AEP; Pal, phosphonoalanine.

The fractions of the total phosphorus in the aminophosphonic acids, estimated from the 88-hr labeling experiment, ranged from $4·4 \times 10^{-5}$ for *Syracosphaera* to $1·4 \times 10^{-4}$ for Amphidinium. This is much lower than the 1–13 per cent of the total phosphorus found in aminophosphonic acids in other organisms. Two factors which will influence the ratio of C—P phosphorus to total phosphorus are: (1) many phytoplankters will scavange and store phosphate for later metabolic demands and (2) 2-AEP is synthesized primarily during cell division. It has been demonstrated that *Tetrahymena pyriformis* converts phosphate into 2-AEP only during the log phase of growth (Rosenberg, 1964), and recent work with synchronous cultures of *Tetrahymena* has demonstrated that the peak period of synthesis of 2-AEP coincides with the maximum index of cell division (Horiguchi *et al.*, in preparation).

We have also cultured a marine diatom, *Cyclotella nana*, two marine yeasts, *Cryptococcus albidus* and *Rhodotorula* sp., a marine bacteria, *Serratia marinorubra*, with ^{32}P-orthophosphate. We were not able to detect the synthesis of aminophosphonic acids in any of these organisms, nor were we able to isolate 2-AEP from two species of brown algae, *Fucus vesiculosus* and *Nereocystis* sp.

DISCUSSION

This is the second demonstration of the biosynthesis of aminophosphonic acids by representatives of photosynthetic phytoplankton. Considering the much higher relative amounts of 2-AEP found in the planktonic amphipod and the microplankton, we must assume that either the fraction of the total phosphorus incorporated into aminophosphonic acids by phytoplankton in nature is higher than found in culture, or that the zooplankton selectively concentrate the aminophosphonic acids obtained in their food. We would expect the zooplankton species to be able to incorporate the preformed aminophosphonic acids into their tissue phospholipids, as has been demonstrated for vertebrates and the housefly, *Musca domestica* (Kandatsu *et al.*, 1965; Kandatsu & Horiguchi, 1965; Bridges & Rickets, 1966). The acids, 2-AEP, *N*-trimethyl-2-AEP and phosphonoalanine, may be considered as analogs of ethanolamine phosphate, choline phosphate and serine phosphate and thus they may replace the common nitrogenous bases of phospholipids.

The fertility of the oceans is primarily a function of the rate of return of utilizable nitrogen and phosphorus to the photosynthetic zone, proceeding primarily by bacterial degradation and upwelling. The present evidence indicates the necessity of considering a further complexity in the cycle of phosphorus in the marine environment. The chemical stability of the C—P bond (65 kcal/mole, Hudson, 1964) suggests that aminophosphonic acids may comprise a significant fraction of the dissolved organic phosphorus in the ocean. An interesting alternative would be the demonstration of the utilization of dissolved aminophosphonic acids by marine plankton. While there is no evidence for the existence of enzymes that can split the C—P bond, several species of bacteria are able to utilize phosphonic acids as a source of metabolic phosphorus (Zeleznick *et al.*, 1963; Mastalerz *et al.*, 1965; Harkness, 1966). The catabolism probably proceeds through transamination and oxidative breakdown of the carbon chain (Roberts *et al.*, 1968; La Nauze & Rosenberg, 1967). This would not be expected to proceed as rapidly as

134

the degradation of ester phosphates by bacteria. Bacteria may not assimilate aminophosphonic acids in the presence of available phosphate. In the one case examined, the bacteria, *Bacillus cereus*, did not utilize 2-AEP until all of the orthophosphate was depleted (Rosenberg & La Nauze, 1967).

Acknowledgements—We appreciate the generous help of D. W. Hood and his staff at the Institute of Marine Science, University of Alaska, in the initial phase of this research and J. L. Barnard, Division of Marine Invertebrates, United States National Museum, Smithsonian Institute, for the identification of *Amonyx nugax*. We also thank R. W. Holmes, Biology Department, University of California at Santa Barbara, for the axenic phytoplankton cultures.

This research was supported by the Office of Naval Research grant NONR-3001-[00]-NR 108-458 to Eugene Roberts (J. S. K. and M. H.) and the Atomic Energy Commission, AEC Contract No. AT (11-1) GEN 10, Project Agreement 20 to J. D. H. Strickland (P. M. W.).

REFERENCES

Bridges R. G. & Ricketts J. (1966) Formation of a phosphonolipid by larvae of the housefly, *Musca domestica. Nature, Lond.* **211**, 199–200.

Harkness R. D. (1966) Bacterial growth on aminoalkylphosphonic acids, *J. Bacteriol.* **92**, 623–627.

Horiguchi M. (1967) C—P compounds. *Tanpakushitsu-kakusan-koso* **12**, 315–323.

Horiguchi M. & Kandatsu M. (1959) Isolation of 2-aminoethane phosphonic acid from rumen protozoa. *Nature, Lond.* **184**, 901–902.

Horiguchi M., Kittredge J. S. & Roberts E. (1968) Biosynthesis of 2-aminoethylphosphonic acid in *Tetrahymena. Biochim. biophys. Acta* **165**, 164–166.

Hudson F. (1964) The nature of the chemical bonding in organophosphorus compounds. *Pure Appl. Chem.* **9**, 371–386.

Kandatsu M., & Horiguchi M. (1965) The occurrence of ciliatine (2-aminoethylphosphonic acid) in the goat liver. *Agr. Biol. Chem.* **29**, 781–782.

Kandatsu M., Horiguchi M. & Tamari M. (1965) The incorporation of ciliatine (2-aminoethylphosphonic acid) into lipids of the rat liver. *Agr. Biol. Chem.* **29**, 779–780.

Kittredge J. S. & Hughes R. R. (1964) The occurrence of α-amino-β-phosphonopropionic acid in the zoanthid, *Zoanthus sociatus*, and the ciliate, *Tetrahymena pyriformis. Biochemistry* **3**, 991–996.

Kittredge J. S., Isbell A. F. & Hughes R. R. (1967) Isolation and characterization of the *N*-methyl derivatives of 2-aminoethylphosphonic acid from the sea anemone, *Anthopleura xanthogrammica. Biochemistry* **6**, 289–295.

Kittredge J. S., Roberts E. & Simonsen D. G. (1962) The occurrence of free 2-aminoethylphosphonic acid in the sea anemone, *Anthopleura elegantissima. Biochemistry* **1**, 624–628.

La Nauze J. M. & Rosenberg H. (1967) The breakdown of aminoethylphosphonate by cell-free extracts of *Bacillus cereus. Biochim. biophys. Acta* **148**, 811–813.

Mastalerz P., Wieczorek Z. & Kochman M. (1965) Utilization of carbon-bound phosphorus by microorganisms. *Acta biochim. polon.* **12**, 151–156.

Quin L. D. (1967) The natural occurrence of compounds with the carbon–phosphorus bond. In *Topics in Phosphorus Chemistry* (Edited by Grayson M. & Griffith E. J.), pp. 23–47. Interscience, New York.

Roberts E., Simonsen D. G., Horiguchi M. & Kittredge J. S. (1968) Transamination of aminoalkylphosphonic acids with α-ketoglutarate. *Science* **159**, 886–888.

Rosenberg H. (1964) The distribution and fate of 2-aminoethylphosphonic acid in *Tetrahymena. Nature, Lond.* 299–300.

ROSENBERG H. & LA NAUZE J. M. (1967) The metabolism of phosphonates by micro-organisms. The transport of aminoethylphosphonic acid in *Bacillus cereus*. *Biochim. biophys. Acta* **141**, 79–90.

VON STOSCH H. A. & DREBS G. (1964) Entwicklungsgeschichtliche Untersuchungen an zentrischen Diatomen—IV. Dir Planktondiatome, *Stephanopyxis turris*: Ihre Behandlung und Entwicklungsgeschichte. *Helgol. wiss. Meersunters.* **11**, 209–257.

ZELEZNICK L. D., MYERS T. C. & TITCHNER E. B. (1963) Growth of *Escherichia coli* on methyl- and ethylphosphonic acids. *Biochim. biophys. Acta* **78**, 564–567.

DDT Reduces Photosynthesis by Marine Phytoplankton

CHARLES F. WURSTER, JR.

Recently it has become apparent that DDT (*1*) and its derivatives are among the most widely distributed synthetic chemicals on Earth. They are found not only in soils (*2*), runoff water (*3*), air, and rainwater (*4*), but also in most animals analyzed from widely diverse parts of the world, including Antarctica (*5*, *6*). Residues of DDT were recently reported in marine organisms of both the Atlantic and Pacific oceans, including pelagic birds at the top of wholly oceanic food chains (*7*); this and other evidence suggest widespread contamination of marine plankton by these chemicals. While components of the zooplankton have been shown to be highly susceptible to DDT (*8*, *9*), very little is known of its effects on phytoplankton. Since a substantial part of the world's photosynthesis is performed by phytoplankton (*10*), interference with this process could be important to the biosphere.

In order to determine the effects of DDT on photosynthesis by marine phytoplankton, species important as food organisms, representing four classes of phytoplankton, were chosen from

137

laboratory cultures maintained at Woods Hole Oceanographic Institution (WHOI). These included the diatom *Skeletonema costatum*, isolated from Long Island Sound (WHOI clone "Skel"); the coccolithophore *Coccolithus huxleyi* from the Sargasso Sea (32°10′N,64°30′W; "BT-6"); the green alga *Pyramimonas* sp. from the Sargasso Sea (33°11′N,65°15′W; "13-10 Pyr"); and the neritic dinoflagellate *Peridinium trochoideum* ("Peri"). In addition, water from Vineyard Sound (Woods Hole) was tested, containing a typical neritic phytoplankton community dominated by various diatoms.

Fluorescent lights (5000 lux) on a cycle of 14 hours light and 10 hours dark were used at 17°C. The four laboratory cultures were grown axenically in half-strength medium "f" (*11*), an enriched sea water. In 125-ml erlenmeyer flasks, 50-ml portions of medium were inoculated to yield the appropriate cell concentrations, and these were cultured for 24 hours. This initial culture period was omitted with *Pyramimonas* and the Vineyard Sound water; the latter was filtered through 50-μ mesh to remove zooplankton and was enriched to "f/100." To each flask 1 or 2 μl of pure p,p'-DDT (*1*) in an ethanolic solution was then added to yield the desired concentrations of DDT. Control flasks received equal amounts of pure ethanol, with no detectable effect on results. After culturing for 20 to 24 more hours, ^{14}C-bicarbonate (*12*) was added, and the algae were illuminated for 4 to 5 more hours. A few controls were run in darkness. Cultures were then collected on 0.8-μ Millipore filters, by use of slightly reduced pressure, and washed three times with filtered sea water; the filtered cultures were dried and radiation was counted. The radioactivity retained by the filtered cells is related to the amount of carbon fixed by photosynthesis (*13*).

Cell concentrations were chosen to approximate densities found in nature.

To facilitate qualitative comparisons between the four phytoplankton species, cell concentrations were calculated to yield approximately equal total cell areas at the end of the experiment. Cell area roughly correlates with metabolic activity in phytoplankton (*14*).

The data from these experiments are shown in Fig. 1; dark uptake of ^{14}C was subtracted from all results. Probabilities (*P*) that a negative linear regression was random ranged from .0007 to .06, with $P < .0001$ for these five experiments combined. Data from DDT concentrations greater than 100 parts per billion (ppb) have been omitted from the regression calculations since they are not on the linear part of the curve and may be subject to anomalies caused by the limited solubility (1.2 ppb) of DDT in water (*15*). The fact that DDT could be immobilized as a solid when the solubility limit was so greatly exceeded may explain the slightly increased uptake of ^{14}C in some instances at 200 to 500 ppb of DDT.

Presumably the effect on phytoplankton occurs after absorption of DDT by the cells, a process to be expected from the very low solubility of DDT in water and its higher solubility in lipid-containing biological material. One might therefore expect increased effect with decreasing cell concentration, caused by an increased amount of DDT per cell. Accordingly in one experiment the concentration of DDT was held constant at various cell concentrations. The validity of the hypothesis is shown by the results of 10 ppb of DDT (Fig. 2): photosynthesis by *Skeletonema* decreased to half that by the controls as the cell concentration was reduced by two orders of magnitude.

The fact that these data apparently follow sigmoid curves is typical of dose-response relations (*16*) and suggests the absence of a threshold concentration of DDT below which no effects occur. Experimental scatter produced some up-

138

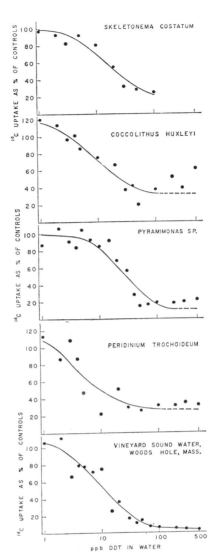

Fig. 1. Photosynthesis by phytoplankton at various concentrations of DDT, measured by uptake of ^{14}C (percentages) relative to uptake by controls. The following are, respectively, calculated cell concentrations (per milliliter) at the end of each experiment, mean counts per minute (cpm) for controls, uptake in the dark (percentages) relative to uptake by controls, and probabilities that a negative linear regression is random: *Skeletonema*, 1600, 1390 cpm, 1.1 percent, $P = .052$; *Coccolithus*, 4920, 1435 cpm, 39.7 percent, $P = .00074$; *Pyramimonas*, 2100, 564 cpm, 22.8 percent, $P = .032$; *Peridinium*, 240, 533 cpm, 53.4 percent, $P = .059$; Vineyard Sound water (as obtained), 6200 cpm, 0.6 percent, $P = .0011$.

139

take of ^{14}C above 100 percent at low concentrations of DDT, however. This should not be interpreted as a low-level stimulatory effect, a possibility that cannot be evaluated from these data.

It is known that DDT reduces Hill reaction activity in barley chloroplasts (17). Photosynthesis by two species of algae and by a natural phytoplankton community was diminished by 1000 ppb of DDT (8), but it now appears that at least some phytoplankton species are sensitive to much lower concentrations.

Evaluation of the significance of these findings to phytoplankton communities in nature is not simple, although the near-ubiquity of DDT residues implies a potential for widespread effects. Representative concentrations of DDT residues within natural phytoplankton communities are not generally known.

Water remote from a site of application of DDT usually averages less than 1 ppb (3), but concentrations in water may be misleading. Natural waters carry DDT in suspension, as well as particulates to which DDT is adsorbed at concentrations 10^4 to 10^5 times those in the water (6). In my experiments half the DDT may be lost by codistillation within 24 hours (18), so that DDT concentrations were probably much lower than is indicated (Figs. 1 and 2) when uptake of ^{14}C was measured.

Absorption of DDT by the cells further reduces the concentration in the water. In nature, concentrations in water are likely to be more constant, the DDT being replaced from persistent residues in mud, detritus, runoff water, and other sources as it is absorbed by cells. Thus a given concentration in natural waters would be of greater biological importance than the same initial concentration in vitro under the conditions of these experiments.

Water near a direct application of DDT to the environment, however, commonly contains concentrations com-

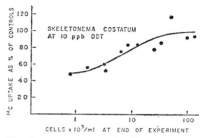

Fig. 2. Photosynthesis in *Skeletonema* at 10 ppb of DDT at various cell concentrations; $P = .032$.

parable to those employed by me. Treatment with DDT of a Florida salt marsh yielded water containing up to 133 ppb for 1 week (19), some California waters contained as much as 22 ppb (6), and 14 to 20 ppb of DDD was added directly to Clear Lake, California (20). Under such circumstances, and especially where cell concentrations are low, photosynthesis and growth by phytoplankton (21) are probably reduced, so that the base of the food chain is diminished and higher organisms are affected.

Selective toxic stress by DDT on certain algae may alter the species composition of a natural phytoplankton community (21). This floral imbalance could favor species normally suppressed by others, producing population explosions and dominance of the community by one or a few species. This process would aggravate the problems of eutrophication caused by over-fertilization from sewage, agricultural runoff, and other artificial sources; it may partially explain the appearance of algal blooms during recent years within certain closed bodies of water. Such effects are insidious and their cause may be obscure, yet they may be ecologically more important than the obvious, direct mortality of larger organisms that is so often reported.

140

References and Notes

1. 1,1,1-Trichloro-2,2-bis(*p*-chlorophenyl)ethane.
2. C. A. Edwards, *Residue Rev.* 13, 83 (1966); H. Cole, D. Barry, D. E. H. Frear, A. Bradford, *Bull. Environ. Contam. Toxicol.* 2, 127 (1967); G. M. Woodwell and F. T. Martin, *Science* 145, 481 (1964).
3. L. Weaver, C. G. Gunnerson, A. W. Breidenbach, J. J. Lichtenberg, *Public Health Rep.* 80, 481 (1965).
4. D. C. Abbott, R. B. Harrison, J. O'G. Tatton, J. Thomson. *Nature* 208, 1317 (1965); *ibid.* 211, 259 (1966); G. A. Wheatley and J. A. Hardman, *ibid.* 207, 486 (1965); P. Antommaria, M. Corn, L. DeMaio, *Science* 150, 1476 (1965).
5. G. M. Woodwell, C. F. Wurster, P. A. Isaacson, *Science* 156, 821 (1967); N. W. Moore and J. O'G. Tatton, *Nature* 207, 42 (1965); W. J. L. Sladen, C. M. Menzie, W. L. Reichel, *ibid.* 210, 670 (1966); J. Robinson, A. Richardson, A. N. Crabtree, J. C. Coulson, G. R. Potts, *ibid.* 214, 1307 (1967); P. A. Butler, in *Trans. North American Wildlife Natural Resources Conf. 31st 14-16 March 1966*, pp. 184-9.
6. J. O. Keith and E. G. Hunt, in *Trans. North American Wildlife Natural Resources Conf. 31st 14-16 March 1966*, pp. 150-77.
7. R. W. Risebrough, D. B. Menzel, D. J. Martin, H. S. Olcott, *Nature* 216, 589 (1967); C. F. Wurster and D. B. Wingate, *Science* 159, 979 (1968).
8. P. A. Butler, *U.S. Fish Wildlife Serv. Circ.* 167 (1963), pp. 12-20.
9. V. L. Loosanoff, J. E. Hanks. A. E. Ganaros, *Science* 125, 1092 (1957); D. S.
Grosch, *ibid.* 155, 592 (1967); H. O. Sanders and O. B. Cope, *Trans. Amer. Fisheries Soc.* 95, 165 (1966).
10. C. S. Yentsch, *Oceanog. Marine Biol. Ann. Rev.* 1, 157 (1963).
11. R. R. L. Guillard and J. H. Ryther, *Can. J. Microbiol.* 8, 229 (1962).
12. Comprising 0.1 to 0.5 ml of a sodium ^{14}C-bicarbonate solution, 10 μc/ml, pH 9.5; New England Nuclear Corp., Boston, Mass.
13. E. S. Nielsen, *J. Conseil Conseil Perm. Intern. Exploration Mer* 18, 117 (1952); M. S. Doty and M. Oguri, *Pubbl. Staz. Zool. Napoli Suppl.* 31, 70 (1959).
14. T. J. Smayda, *Inter-Amer. Trop. Tuna Comm. Bull.* 9, 467 (1965).
15. M. C. Bowman, F. Acree, M. K. Corbett, *J. Agr. Food Chem.* 8, 406 (1960).
16. W. W. Westerfeld, *Science* 123, 1017 (1956).
17. P. D. Lawler and L. J. Rogers, *Nature* 215, 1515 (1967).
18. F. Acree, M. Beroza, M. C. Bowman, *J. Agr. Food Chem.* 11, 278 (1963).
19. R. A. Croker and A. J. Wilson, *Trans. Amer. Fisheries Soc.* 94, 152 (1965).
20. A. I. Bischoff, *Calif. Fish Game* 46, 91 (1960).
21. N. Lazaroff and R. B. Moore, *J. Phycol. Suppl.* 2, 7 (1966); E. L. Bishop, *Public Health Rep.* 62, 1263 (1947).
22. Aided by the Research Foundation of the State University of New York. I thank the Woods Hole Oceanographic Institution where this work was performed, the Marine Biological Laboratory, and many individuals, especially R. R. L. Guillard, C. S. Yentsch, J. A. Hellebust, L. B. Slobodkin, H. I. Stanley, and C. H. Spielvogel, for assistance.

DDT Residues in Marine Phytoplankton: Increase from 1955 to 1969

JAMES L. COX

Annual use of DDT in the United States has declined in the past decade (*1*), yet there is recent evidence of abnormally high DDT residues in marine fish from U.S. coastal waters, and such contamination in these are may exceed that in freshwater habit..is (*2*). This could indicate either (i) that environmental DDT residues are increasing or (ii) that these recent analyses simply reflect current DDT input and that DDT concentrations have in fact been even higher in the past. Although DDT residues in estuarine shellfish (*3*) have shown no consistent upward or downward trends, the time has been too short and the estuarine system too responsive to weather conditions and local sources of pesticides to provide any measure of the trends in the coastal environment. Declining reproductive success in species of marine pelagic birds, attributable to DDT residues (*4*), does suggest that residues of DDT are increasing in the coastal pelagic food chains of which these birds are high-order consumers.

A decision between the alternatives could be made if historical collections of marine organisms were available. At the Hopkins Marine Station, samples (composed primarily of phytoplankton) collected with a fine-mesh net from Monterey Bay, California, have been collected from 1955 to 1969 (*5*). Phytoplankton samples are particularly suited for analysis because they represent the first link in pelagic food chains. Trends in their concentrations of DDT residues are relevant

to all higher-order consumers on the food chain. Also, DDT uptake by phytoplankton is rapid and essentially irreversible (*6*); thus, it can be assumed that the content of DDT residues of phytoplankton reflect prevailing amounts of environmental DDT. To examine the change in content of DDT residues over the collection period, 23 samples from the collection were analyzed. All the samples had been preserved in a 3 percent solution of formalin in seawater. The estimated concentrations of DDT residues (*7*) for the samples were based on their carbon content as determined by wet combustion (*8*) of replicate portions. Formalin induces error in carbon determinations of marine planktonic material (*9*), but the errors in this instance were small (< 10 percent). Treatment of freshly collected material from the same station with formalin had no apparent effect on estimates of the DDT content when compared to that of frozen controls.

Samples were filtered onto combusted GFC glass-fiber filters (Whatman) after filtration through 0.33-mm netting to remove larger zooplankton. The sample and filter pad were ground together in three successive rinses of high-purity *n*-hexane. The pooled rinses were concentrated and chromatographed on silica-gel microcolumns (*10*). Eluates from the columns were concentrated at 37°C under a stream of nitrogen and analyzed by gas-liquid chromatography (GLC). All glassware used in the procedure was combusted at 350°C overnight prior to use; this treatment reduced background contamination nearly to zero for the GLC analyses. Recovery from a variety of samples of known content of DDT exceeded 95 percent in all cases.

The extracts were injected into a Beckman GC-4 gas chromatograph equipped with two columns and two electron-capture detectors (11). Each sample was chromatographed on at least two columns of different composition (12). Peaks were identified by standard injection retention times, fortified injections, and disappearance of presumptive DDT and DDD peaks caused by treatment of the extracts with KOH in alcohol.

There were higher concentrations of DDT residues in more recent samples (Fig. 1). Inasmuch as sample storage may have affected this apparent temporal trend, experiments were performed to test the effect of decomposition on the relative proportions of the three DDT constituents found in the samples. Ring-labeled ^{14}C-DDT was added to sealed ampules that contained portions of a phytoplankton sample preserved in formalin. The amounts of labeled compound added were comparable to those amounts found in the 23 analyzed samples. Because the samples had been stored in the dark, it was assumed that any possible breakdown would be thermochemical, not photochemical. Elevated temperatures for short periods of time were used to recreate longer periods at room temperature. The contents were heated at 30°C, 60°C, and 75°C for 6 days and then removed from the ampules, extracted, and analyzed by thin-layer chromatography (13). Narrow (0.5-cm) zones were scraped from the chromatoplates into scintillation vials for measurements of ^{14}C activity. The degree to which DDT was broken down to polar compounds affected the relative proportions of the remaining nonpolar constituents: p,p'-DDT, p,p'-DDD, and p,p'-DDE. No breakdown of the ^{14}C-DDT occurred in the samples heated at

30°C. In the samples heated at 60°C, 28 percent of the ^{14}C-DDT broke down to polar compounds, but about half of the remaining nonpolar material was p,p'-DDT. In the sample heated at 75°C, 38 percent of the ^{14}C-DDT broke down to the polar compounds, but p,p'-DDT comprised only 15 percent of the nonpolar material, whereas p,p'-DDD and p,p'-DDE comprised 83 percent. On the basis of these experiments, a change in the relative proportions of the DDT residues in a sample would be expected if any net decomposition to nonpolar products had occurred during storage. In fact, the relative proportions of the DDT constituents found in the samples were quite constant (14); percent regression analysis showed the slope of each percentage versus time function to be not significantly different from zero. Therefore, the trend indicated in Fig. 1 represents an actual increase in the DDT residues in the phytoplankton rather than a loss of analyzable residues during sample storage.

Part of the variability in the values in Fig. 1 is attributable to sample size. Due to the nature of the collection technique, the sample sizes were directly related to the density of the phytoplankton in the water at the time of collection. The estimates of carbon content in the 23 samples were based on standard portions of the samples, which in turn contained the entire contents of a vertical ¼-meter net tow from a depth of 15 meters to the surface. The carbon values are thus indices of standing crop density (Fig. 2). The assumption behind the theoretical curve (Fig. 2) is that a fixed amount of pesticide residue becomes incorporated in the algal material present in a given volume of water, regardless of the density of the standing crop. However, density of the standing crop affects the final concentration of acquired residues according to the relationship in Fig. 2. This suggests that the partition coefficient of DDT residues for phytoplankton and similar ma-

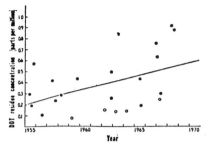

Fig. 1. Concentrations of DDT residues
(7) in samples of phytoplankton collected
by towed nets from Monterey Bay, California, 1955 to 1969. Concentrations are
expressed as weight of estimated DDT
residues per unit wet weight of phytoplankton as converted from measurements of
oxidizable organic carbon content of the
samples (16). Solid circles indicate small-
est samples (<0. mg of carbon); half-
solid circles indicate samples with 0.2 to
0.4 mg of carbon; open circles indicate
samples with greater than 0.4 mg of carbon.

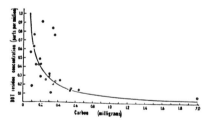

Fig. 2. The effect of size of relative standing crop (milligrams of carbon) on the
estimated concentration of DDT residue
(7). The theoretical curve was computed
according to the relationship $C \times D = k$,
where C = carbon content, D = DDT
residue concentration and k = weight of
the mean amount of DDT residues in the
samples. Solid circles indicate samples
taken from the later half of the sampling
period; open circles indicate samples from
the earlier half. Values on the vertical
axis were derived as in Fig. 1.

terial (*6, 15*) diminishes as the density of the phytoplankton increases. A comparison of the points falling above or below the theoretical curve in Fig. 2 shows that the preponderance of later points have higher concentration values, despite the effect of the size of standing crop. The same conclusion may be reached by examining the points in Fig. 1 by size classes. The residues of DDT may be increasing in the primary stages of coastal pelagic food chains. If the processes of decomposition and dispersal of these residues in succeeding steps are not sufficiently rapid to counteract this apparent increase, a delay may be expected before the decline of domestic usage of DDT begins to be reflected in the components of these food chains.

References and Notes

1. T. Eichers and R. Jenkins, *U.S. Dep. Agr. Agr. Econ. Rep. No. 158* (1969).
2. R. W. Risebrough, *Chemical Fallout, First Rochester Conference on Toxicity* (Thomas, Springeld, Ill., 1969), p. 8; also analyses contracted by the California Department of Public Health showed some samples of canned jack mackerel *Trachurus symmetricus*, Ayres, from California waters had in excess of 5.0 parts of DDT residues per million.
3. P. A. Butler, *Bioscience* 19, 889 (1969); *U.S. Bur. Comm. Fish. Res. Contract Rep. 14-17-0002-265* (1970).
4. R. W. Risebrough, D. B. Menzel, D. J. Martin, Jr., H. S. Olcott, *Nature* 216, 589 (1967); C. F. Wurster, Jr., and D. B. Wingate, *Science* 159, 979 (1968); D. B. Peakall, *Sci. Amer.* 222, 72 (1970); D. B. Peakall, *Science* 168, 592 (1970).
5. The samples discussed here are part of a larger series which has been described in detail by R. L. Bolin and D. P. Abbott, *Calif. Coop. Ocean. Fish. Invest. Rep.* 9, 23 (1963) and D. P. Abbott and R. Albee, *ibid.* 11, 55 (1967).
6. A. Sodergren, *Oikos* 19, 126 (1967); J. Cox, *Bull. Environ. Contam. Toxicol.* 5, 218 (1970).
7. The DDT residues include all the constituents of technical DDT and the nonpolar metabolites derived therefrom. In this study only p,p'-DDT [1,1,1-trichloro-2,2-bis(p-chlorophenyl)ethane],p,p'-DDD [1,1-dichloro-2,2-bis(p-chlorophenyl)ethane], and p,p-DDE [1,1-dichloro-2,2-bis(p-chlorophenyl)ethylene] were detected in measurable amounts. The term DDT residues as it is used in the text refers to all of these three compounds.
8. J. D. H. Strickland and T. R. Parsons, *Bull. Fish. Res. Board Can.* 167, 207 (1968).
9. T. L. Hopkins, *J. Cons. Cons. Perma. Int. Explor. Mer* 31, 300 (1968).
10. A. M. Kadoum, *Bull. Environ. Contam. Toxicol.* 3, 65 (1968); *ibid.*, p. 354.
11. All GLC parameters were those suggested in *Pesticide Analytical Manual* (U.S. Dept. of Health, Education, and Welfare, Food and Drug Administration, revised, 1968), vol. 2
12. Coatings used on the columns were 5 percent DC-200, 5 percent QF-1, 5 percent mixed bed of DC-200 and QF-1, and 3 percent SE-30 with 6 percent QF-1 in a mixed bed. All coatings were made on DMCS Chromosorb W.
13. Silica gel G was used as the adsorbent. Chromatoplates were developed in *n*-heptane, compounds were identified by cochromatography with pure standards.
14. Values expressed as percent followed by standard error in percent: p,p'-DDT, 57.1 ± 12.9; p,p'-DDE, 18.0 ± 6.7; and p,p'-DDD, 24.9 ± 13.7.
15. C. F. Wurster, Jr., *Science* 159, 1474 (1968).
16. Wet weight concentrations were computed by converting carbon content of the samples to estimated wet weights by multiplying by 100. This factor was derived from values given by E. Harris and G. A. Riley [*Bull. Bingham Oceanogr. Collect. Yale Univ.* 15, 315 (1956)] and H. Curl, Jr. [*J. Mar. Res.* 20, 181 (1962)].
17. Supported by NSF grant GB 8408, a grant from the State of California Marine Research Committee, and an NSF predoctoral fellowship. I thank David Bracher for technical assistance with the carbon and DDT analyses.

Oxygen-18 Studies of Recent Planktonic Foraminifera:

Comparisons of Phenotypes and of Test Parts

Alan D. Hecht
Samuel M. Savin

We report herein comparisons of oxygen isotopes in phenotypes of the same species of Recent Planktonic Foraminifera and in test parts of individual species. We have studied the following paleontologic problems by means of paleotemperature techniques: (i) the significance of the presence or absence of diminutive final chambers (*1*); (ii) the significance of the presence or absence of bullae (platelike structures covering the primary and secondary apertures); (iii) the significance of the shape of the final chamber in *Globogerinoides sacculifer-trilobus*; and (iv) the validity of Bé's hypothesis that *Spheroidinella dehiscens* is an aberrant deep-water form of *G. sacculifer-trilobus* (*2*).

All samples for isotopic analysis, taken from core tops except where otherwise indicated, were washed in distilled water and sized. Hand-picked samples from the fraction larger than 250 μm were reacted uncrushed and isotopically analyzed by means of standard techniques (*3*). Values of δ (*4*) are reported relative to the PDB-I (Pee Dee Formation belemnite) standard. Isotopic temperatures were calculated by means of the equation given by Craig (*5*).

Isotopic compositions of seawater were taken from the literature for surface stations as close as possible to the core locations (*6, 7*). Estimates of the depths at which the Foraminifera populations occurred were made from isotopic temperatures and bathythermographic data supplied by the National Oceanographic Data Center. All results are shown in Table 1.

Berger (*1*) has suggested that the presence of diminutive final chambers in Planktonic Foraminifera from ocean sediments is indicative of growth in a stressed environment in the overlying water column. This stress may be physical (temperature, salinity) or biological (insufficient food). Our data show that phenotypes of *Globogerinoides ruber* with a diminutive final chamber from three localities have isotopic temperatures 1° to 4.5°C colder than do phenotypes with a normal final chamber from the same samples. The temperature difference is greatest in the sample from the Blake Plateau, in which the small final chamber of the test was also flattened. No significant difference in temperatures was obtained for pink and white "normal" types from one locality. If a species has a range of optimum depth at or near the surface, as does *G. ruber*, then a stressed environment in the same water column must lie in deeper, colder water. Thus, the isotopic data for this species are compatible with Berger's hypothesis.

In the case of species with distributions of optimum depth at intermediate depths, a stressed environment might lie either in cooler, deeper water; in warmer, shallower water; or conceivably in both simultaneously. *Globorotalia cultrata* lives at intermediate depths. Two samples of this species showed that the phenotype with a normal final chamber recorded essentially the same temperature as the phenotype with a diminutive final chamber. In a third sample, from an area of coastal upwelling, the "normal" phenotype recorded a temperature about 1.5°C colder. In the case of *Globoquadrina dutertrei* (8), another species living at intermediate depths, both phenotypes of one sample recorded virtually identical isotopic temperatures. In another sample, the form with a diminutive final chamber recorded a temperature 1.2°C colder than did the "normal" phenotype, but the difference is virtually at the level of uncertainty for these analyses. Thus our results suggest that the presence of a diminutive final chamber may be related to some environmental factors in that phenotypes may have different distributions in the water column, but that temperature is not the only factor in determining the occurrence of this phenotype.

In the case of *Globogerinoides conglobatus,* we have investigated the possible existence of temperature differences between populations with and without bullae. A bullate population in the Gulf of Mexico recorded a temperature 1.8°C warmer than the nonbullate one did. Three other samples showed no significant differences in temperatures between phenotypes.

Globogerinoides sacculifer and *G. trilobus* have been distinguished from each other on the basis of a saclike as opposed to a more spherical final chamber. According to Bermudez (9), the final chamber has the function of increasing the buoyancy of the test; thus the sacculifer form would live in shallower water than would the trilobus type. However, Jones (10) has observed *G. sacculifer* at greater depths than *G. trilobus* in plankton tows in the Carribbean area.

Our results show that *G. sacculifer* and *G. trilobus* are isotopically similar. In samples from the Gulf of Mexico and the Indian Ocean *G. sacculifer* recorded a slightly warmer temperature than *G. trilobus,* whereas for a few Atlantic samples, *G. trilobus* recorded a slightly warmer temperature. Isotopic comparison of the last chamber of *G. sacculifer* with the entire test for a sample from the Gulf of Mexico shows the two to be almost identical.

The data indicate that both *G. sacculifer* and *G. trilobus* form their tests at similar temperatures, and presumably at similar depths. Thus, depth stratification at any one place as reported by Jones may be related to factors other than temperature. Our results do not exclude the possibility that the final chamber in *G. sacculifer* increases the buoyancy of the test, but it does suggest that this chamber is formed at the same temperature as the bulk of the early formed parts of the test.

Finally, Bé (2) has suggested that *Spheroidinella dehiscens* is an aberrant terminal form of *G. sacculifer-trilobus* which, after early growth at shallow depths, sinks to bathypelagic depths (300 to 2000 m), where maximum encrustation with additional calcite occurs. However, other workers argue that *S. dehiscens* is a separate species which may live in waters as shallow as 150 m (11).

Our isotopic comparisons of the outer crust of *S. dehiscens* (12), the entire test of the same species, and *G. sacculifer-trilobus* from the same sample population show that in an equatorial Atlantic sample the crust records a warmer temperature than the entire

147

Table 1. Results of oxygen isotope analyses.

Core name	Location	Species	Phenotype	δO¹⁸ (‰)	Runs (No.)	Average deviation (‰)	δ-water (‰)	Isotopic temperature (°C)	Average depth (m)*	Depth range (m)*
			Atlantic							
Blake Plateau	32°53'N	G. ruber	Normal chamber	-1.55	2	.01	+0.87†	27.8	10	0-20
Blake Plateau	77°18'W	G. ruber	Diminutive chamber	-0.77	2	.07		23.3	95	10-120
			Gulf of Mexico							
G-1290	25°35'N	G. ruber	Normal chamber pink	-1.76	2	.10	+0.68†	27.9	15	0-35
G-1290	84°49'W	G. ruber	Normal chamber white	-1.67	3	.15		27.5	15	0-50
G-1290	84°49'W	G. ruber	Diminutive chamber	-1.45	2	.18		26.4	30	0-60
			Pacific							
DWHG 84	15°S	G. ruber	Normal chamber	-0.59	2	.03	+0.22‡	20.6	65	55-85
DWHG 84	112°W	G. ruber	Diminutive chamber	-0.07	2	.03		18.3	75	65-95
			Atlantic							
C-1	2°48'N	G. trilobus		-0.78	2	.10	+0.43‡	22.1	60	50-70
C-1	16°28'W	G. sacculifer		-0.48	2	.09		20.8	65	65-75
V22-28	5°49'N	G. trilobus		-1.16	2	.14	+0.43‡	23.9	50	40-60
V22-28	39°16'W	G. sacculifer		-0.91	2	.14		22.8	60	55-70
A181-7	10°33'N	G. trilobus		-1.09	2	.02	+0.43	23.6	50	40-60
A181-7	57°20'W	G. sacculifer		-1.08	2	.09		23.5	50	40-60
Blake Plateau	32°53'N	G. trilobus		.43	2	.18	+0.87†	22.6	100	85-150
Blake Plateau	77°18'W	G. sacculifer		.44	2	.11		22.6	100	85-150
			Gulf of Mexico							
G-1290	25°35'N	G. trilobus	Whole	.86	2	.08	+0.68†	23.7	60	0-80
G-1290	84°49'W	G. sacculifer		-1.03	2	.09		24.4	50	0-80
G-1290	84°49'W	G. sacculifer	Last chamber	-1.09	2	.02		24.7	50	0-80
			Pacific							
DWHG 84	15°S	G. trilobus		-.25	2	.05	+0.22‡	19.1	70	60-90
DWHG 84	112°W	G. sacculifer		-.38	2	.13		19.7	70	60-90
			Indian Ocean							
DODO 124D	11°54'S	G. trilobus		-1.57	2	.03	+0.19‡	24.7	40	N.A.§
DODO 124D	66°27'E	G. sacculifer		-1.84	2	.15		26.0	30	N.A.§
			Atlantic							
C-1	2°48'N	S. dehiscens	Whole test	+.14	3	.06	+0.43‡	18.1	75	65-85
C-1	16°28'W	S. dehiscens	Outer crust	+0.04	2	.04		18.6	75	65-85

Core	Location	Species	Chamber	Region	δ (warmest/coldest)	N	s.d.	Avg. δ	Temp (°C)	Depth	Depth range
DODO 124D	11°54'S	*S. dehiscens*	Whole	*Indian Ocean*	−1.07	2	.24	+0.19‡	22.4	55	N.A.
DODO 124D	66°27'E	*S. dehiscens*	Outer crust		−0.99	1			22.0	60	N.A.
E-1	33°00'N	*G. conglobatus*	Bulla	*Atlantic*	−.06	2	.07	+0.87†	20.9	N.A.	
E-1	72°00'W	*G. conglobatus*	Non-bulla		−.07	2	.05		21.0		
G-1290	25°35'N	*G. conglobatus*	Bulla	*Gulf of Mexico*	−.42	2	.17	+0.68†	21.7	75	55–105
G-1290	84°49'W	*G. conglobatus*	Non-bulla		−.03	2	.12		19.9	100	70–135
DWHG 84	15°S	*G. conglobatus*	Bulla	*Pacific Ocean*	−.09	2	.02	+0.22‡	18.5	70	60–95
DWHG 84	112°W	*G. conglobatus*	Non-bulla		−.03	2	.01		17.9	75	60–100
DODO 124D	11°54'S	*G. conglobatus*	Bulla	*Indian Ocean*	−1.39	2	.05	+0.19‡	23.9	45	N.A.
DODO 124D	66°27'E	*G. conglobatus*	Non-bulla		−1.47	2	.11		24.5	40	N.A.
C-1	2°48'N	*G. cultrata*	Normal chamber	*Atlantic*	+.72	2	.04	+0.43‡	15.7	100	95–110
C-1	16°28'W	*G. cultrata*	Diminutive chamber		+.37	2	.06		17.2	85	70–95
A-181-7	10°33'N	*G. cultrata*	Normal chamber		+.12	2	.00	+0.43‡	18.2	75	65–95
A-181-7	57°20'W	*G. cultrata*	Diminutive chamber		+.07	2	.05		18.4	75	65–95
G-1290	25°35'N	*G. cultrata*	Normal chamber	*Gulf of Mexico*	−.02	2	.12	+0.68†	19.9	95	70–130
G-1290	84°49'W	*G. cultrata*	Diminutive chamber		−.08	2	.07		20.1	95	70–130
G-1290	25°35'N	*G. dutertrei*	Normal chamber		−.24	2	.11	+0.68†	20.9	80	60–120
G-1290	84°49'W	*G. dutertrei*	Diminutive chamber		+.02	2	.11		19.7	105	70–130
C-1	2°48'N	*G. dutertrei*	Normal chamber	*Atlantic*	+.10	2	.07	+0.43‡	18.4	75	60–85
C-1	16°28'W	*G. dutertrei*	Diminutive chamber		insufficient material						
DWBG 140	5°S	*G. dutertrei*	Normal chamber	*Pacific* ‖	+.37	2	.01	+0.22‡	16.4	85	70–105
DWBG 140	112°W	*G. dutertrei*	Diminutive chamber		+.32	2	.22		16.6	85	70–105

* Average depth is calculated from average depth-temperature profile. Depth range is calculated from warmest and coldest depth-temperature profiles available. † Epstein and Mayaeda (6). ‡ Craig and Gordon (7). § Not available. ‖ Sample from core at level of 2 to 5 cm.

animal does. Temperatures for the entire animal and the crust are both colder than that recorded by either *G. sacculifer* or *G. trilobus* from the same location, so it is inconceivable that *S. dehiscens* from this location is a variant of *G. sacculifer-trilobus* formed by encrustation at depth. The Indian Ocean comparison is not as clear-cut in that the test of *S. dehiscens* has an isotopic composition intermediate between that of its crust and that of *G. sacculifer-trilobus* from the same sample. Visual estimates indicate that the crust comprises roughly 70 ± 20 percent of the mass of *S. dehiscens*. Thus, within the uncertainty limits of the visual estimation of the relative masses of the two parts, the measured isotopic composition of the bulk is not far from the calculated isotopic composition of a mixture of 70 percent crust and 30 percent *G. sacculifer-trilobus*. Although the isotope data do not rule out the possibility that *S. dehiscens* from this location is simply encrusted *G. sacculifer-trilobus*, encrustation could not possibly have taken place at the depths suggested by Bé. The sum of the isotope data argues very strongly for the consideration of *S. dehiscens* as a separate species.

In summary, phenotypes of a single species and test parts of individual phenotypes sometimes record different isotopic temperatures. Where test parts record different temperatures, as in the case of *S. dehiscens*, conclusions may be drawn concerning the temperatures at which the animal lived during different stages of growth. The occurrence of diminutive final chambers is correlated with temperature for the shallow-water species, *G. ruber*. This is particularly important for paleotemper-

ature studies, since if our model is correct, temperatures determined on entire populations of shallow-water species may be colder than those determined when only the "normal" phenotype is used.

References and Notes

1. Our phenotypes with diminutive final chambers are equivalent to Berger's "Kummerform" types. W. H. Berger, *Amer. Ass. Petrol. Geol. Bull.* **53**, 706 (1969).
2. A. W. H. Bé, *Micropaleontology* **11**, 81 (1965).
3. S. Epstein, R. Buchsbaum, H. A. Lowenstam, H. C. Urey, *Bull. Geol. Soc. Amer.* **62**, 417 (1962).
4. δ is defined as parts per mil deviation from a standard:

$$\delta O^{18} = \frac{(O^{18}/O^{16})_{sample} - (O^{18}/O^{16})_{standard}}{(O^{18}/O^{16})_{standard}} \times 1000$$

5. H. Craig, in *Proceedings of the Spoleto Conference on Stable Isotopes in Oceanographic Studies and Paleotemperatures* (Consiglio Nazionale delle Ricerche, Laboratorio di Geologia Nucleare, Pisa, Italy, 1965), vol. 3, pp. 1–24.
6. S. Epstein and T. Mayeda, *Geochim. Cosmochim. Acta* **4**, 213 (1953).
7. H. Craig and L. I. Gordon, in *Proceedings of the Spoleto Conference on Stable Isotopes in Oceanographic Studies and Paleotemperatures* (Consiglio Nazionale delle Ricerche, Laboratorio di Geologia Nucleare, Pisa, Italy, 1965), vol. 2, pp. 1–122.
8. Our normalform type is equivalent to Zobel's *G. dutertrei-eggeri* [B. Zobel, *Geol. Jahrb.* **85**, 97 (1968), figs. 1–3]. Types with diminutive final chamber appear equivalent to Zobel's *G. dutertrei* form D (fig. 4).
9. P. J. Bermudez, *Biol. Geol.* 3 (special vol.), 119 (1961).
10. J. I. Jones, *Micropaleontology* **13**, 489 (1967).
11. F. G. Parker, *Bull. Amer. Paleontol.* **52**, 115 (1967); W. H. Blow, *Proc. Micropaleontol. Colloq. 1st*, Geneva **1**, 199 (1967); J. A. Wilcoxon, *Contrib. Cushman Found. Foraminifera Res.* **15**, 1 (1964).
12. The outer crust was separated from the rest of the test by breaking the test and removing as much inner material as possible.
13. Supported by research grants from the GSA and Sigma Xi to *t*.D.H. and by NSF grant GA-1693. R. Cifelli, W. Ruddiman, and R. G. Douglas provided samples for analysis. One sample was collected by A.D.H. while on board the oceanographic research vessel *Eastward* of Duke University. We thank A. W. H. Bé, R. G. Douglas, J. Hower, and F. G. Stehli for helpful discussions. Contribution No. 63 of the Case Western Reserve University Department of Geology.

DAVID W. MENZEL
JUDITH ANDERSON, ANN RANDTKE

Marine Phytoplankton Vary in Their Response to Chlorinated Hydrocarbons

Inhibition of photosynthesis by DDT in four species of marine phytoplankton, and in a natural phytoplankton community, has been documented (1). The photosynthesis curves were typical of dose-response reactions (2) although, in general, pronounced toxicity occurred at concentrations well in excess of 1.2 parts per billion (ppb), the solubility of DDT in water (3). It is not clear how chlorinated hydrocarbons effect photo- synthesis in unicellular algae, though it may be inferred from previous data (1, 4) that some possess a marked capacity to concentrate these compounds from the aquatic media.

Tests were made to determine whether organisms isolated from markedly dif- ferent oceanic environments vary in their response to three chlorinated in- secticides. Four species in culture were assayed for their response to dieldrin, endrin, and DDT (5) which are identified in that order as the most widely distributed chlorinated hydrocarbons in major U.S. river basins (6). The species assayed included *Skeletonema costatum* (WHOI clone "Skel."), a coastal centric diatom isolated from Long Island Sound; the naked green flagellate *Dunaliella tertiolecta* (WHOI clone "Dun") typical of tide pools and estuaries; the coccolithophorid *Coccolithus huxleyi* (WHOI clone BT-6) and the centric diatom, *Cyclotella nana* (WHOI clone 13-1), both from the Sargasso Sea.

In all experiments the cultures were illuminated by fluorescent lights (6000 lux) and were grown in half-strength medium "f" (7). Cell carbon concentrations were adjusted to 100, 250, and 500 μg of carbon per liter, considered within the range of naturally occurring carbon concentrations in surface oceanic waters (8). Within these limits no effect of cell concentration on toxicity was noted. For short-term dose-response experiments, cultures, in duplicate, were added to 33-ml screw cap pyrex tubes to which were also added varying concentrations (0.01 to 1000 ppb) of the insecticide dissolved in 5 μl of either acetone or ethanol. The same amount of solvent, previously shown not to affect ^{14}C uptake, was added to the control tubes. To each tube 1 μc of [^{14}C]Na$_2$CO$_3$ was also added. After 24 hours' exposure to light the plants were filtered and counted in a Geiger-Müller end-window counter.

Long-term effects of DDT and endrin on cell division were studied by counting cells each day for 7 days. Cultures were inoculated into 125-ml Erlenmeyer flasks that contained 50-ml portions of media to yield cell concentrations of approximately 10^4 cell/ml (about 200 μg of carbon per liter). To each flask 100 ppb of insecticide was added daily, and an equal volume of solvent was added to each of the controls. Counts were based on the average of four flasks.

None of the insecticides tested at any concentration up to 1000 ppb affected cultures of *Dunaliella*. Although there is considerable scatter in the data in the dose-response experiments, trends indicating toxicity were not evident. Furthermore, no effect on the rate of cell division was measured over a 7-day period (Fig. 1). This species is apparently insensitive to the compounds tested, up to 1000 ppb.

The rate of ^{14}C uptake in *Skeletonema* and *Coccolithus*, on the other hand, was reduced significantly (1) at concentrations above 10 ppb by all three insecticides. The DDT at 100 ppb (added each day) blocked cell division after 2 to 3 divisions in *Skeletonema* but had no apparent effect on *Coccolithus*. Endrin, contrarily, had little effect on the final concentration of *Skeletonema* cells, although the rate of growth over the first 5 days was considerably slower than that of the controls. Reduced growth rates occurred throughout the experiment in *Coccolithus* (Fig. 1).

In contrast to the above species, ^{11}C uptake in *Cyclotella* (Fig. 1) was inhibited by all three insecticides at concentrations above 1 ppb. The slopes of the dose-response curves for dieldrin and endrin suggest that these may have been inhibitory as concentrations down to 0.01 ppb. Cell division was completely inhibited by dieldrin and endrin, whereas cells exposed to DDT divided, but more slowly than the controls.

To interpret the ecological significance of observations concerning the toxicity of these insecticides it is important to recall that all have very low water solubilities. For example, water can carry 1.2 ppb of DDT (3) and 100 ppb of dieldrin (9) in solution. Concentrations above these must be accommodated by precipitation or adsorption to surfaces. Some species responded to concentrations above solubility limits, which indicates that they are capable of incorporating the compounds as small particles or that saturation is maintained while they concentrate the pesticide from solu-

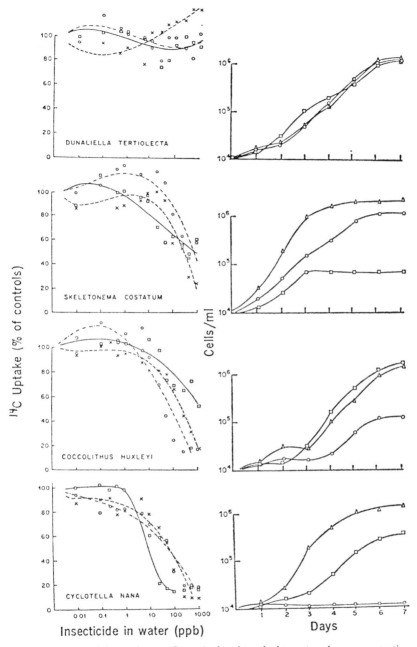

Fig. 1. Left side of figure shows ^{14}C uptake by phytoplankton at various concentrations of DDT (squares and solid lines), dieldrin (crosses and dashed lines), and endrin (open circles and dash-dot lines) relative to uptake by controls (percentages) over 24 hours. The curves were fitted to third-order polynomials by the method of least squares. Right side of the figure shows growth rates of the same species (cells per milliliter) as a function of time when 100 ppb of DDT (squares) and endrin (circles) were added each day for 7 days and solvent was added in equal volume to the controls (triangles).

153

tion. These possibilities were tested by adding 0, 3, 10, 100, and 1000 ppb of DDT to seawater, filtering the solution, and adding *Coccolithus* (200 μg of carbon per liter) to each filtrate. No dose-response reactions characteristic of the above-described experiments were detected; all gave the same ^{14}C uptake in 24 hours as did the controls. Since *Coccolithus* was inhibited at 10 ppb (Fig. 1) the concentration of DDT in all cases must have been reduced to less than that.

Sensitivity and response to environmental pollutants may vary considerably among species of marine planktonic algae. The greater resistance of the one estuarine species, *Dunaliella*, in comparison with the susceptibility of coastal and open-ocean forms like *Skeletonema* and *Coccolithus* and the extreme sensitivity of *Cyclotella*, may reflect the need for adaptability in the face of more unstable conditions close to hand. Although chlorinated hydrocarbons may not be universally toxic to all species, they may exert a dramatic influence on the succession and dominance of individual forms (*10*).

References and Notes

1. C. F. Wurster, Jr., *Science* **159**, 1474 (1968).
2. W. W. Westerfeld, *ibid.* **123**, 1071 (1956).
3. M. C. Bowman, F. Acree, M. K. Corbett, *J. Agr. Food Chem.* **8**, 406 (1960).
4. A. Södergren, *Oikos* **19**, 126 (1968).
5. The DDT is 1,1,1-trichloro-2,2-*bis*(*p*-chlorophenyl) ethane. Dieldrin is 1,2,3,4,10,10-hexachloro-6,7-epoxy-1,4,4a,5,6,7,8,8a-octohydro-1,4-endo-exo-5,8-dimethanonaphthalene. Endrin is 1,2,3,4,10,10-hexachloro-6,7-epoxy-1,4,4a,5,6,7,8,8a-octohydro-1,4-endoendo-5,8-dimethanonaphthalene.
6. L. Weaver, C. G. Gunnerson, A. W. Breidenbach, J. J. Lichtenberg, *Publ. Health Rep. (U.S.)* **80**, 481 (1965).
7. R. R. L. Guillard and J. H. Ryther, *Can. J. Microbiol.* **8**, 229 (1962).
8. D. W. Menzel, *Deep-Sea Res.* **14**, 229 (1967).
9. C. A. Edwards, *Residue Rev.* **13**, 83 (1966).
10. N. Lazaroff, R. B. Moore, *J. Phycol. Suppl.* **2**, 7 (1966).
11. Supported in part by NSF grant GB 15103 and AEC contract AT(30-1)-3862 (Ref. NYO-3862-25). Contribution No. 2424 from the Woods Hole Oceanographic Institution, Woods Hole, Mass. 02543.

Is the World's Oxygen Supply Threatened?

IN an address entitled "Can the World be Saved?" presented in 1967 at the AAAS Meetings in New York[1], Lamont C. Cole expressed concern that the release into the ocean of pesticides and/or other toxic pollutants may, by poisoning the marine diatoms, "bring disaster upon ourselves" in cutting off 70 per cent or more of the world's supply of photosynthetically produced oxygen. This particular doomsday prophecy has gained momentum very significantly in the ensuing 2 years and is now a recurrent theme in almost any discussion of the environment and its abuses by man. This is true not only of dialogue in the popular press and on television but in scientific circles as well. For example, at a recent symposium on primary productivity held at the State University of New York at Albany (March 12–13, 1970) the subject was invoked repeatedly in discussions following papers dealing with natural photosynthetic processes. Clearly, this has become a matter of serious concern. Is it justified?

First, Cole's statement that "70 per cent or more of the total oxygen production by photosynthesis occurs in the ocean and is largely produced by planktonic diatoms" seems to have been based on rather old information. The most recent estimates of the global production of organic matter indicate that the oceans contribute less than one-third to the world total[2].

Second, recent evidence[3] reveals that several chlorinated hydrocarbons exhibit a selective effect on marine phytoplankton, strongly inhibiting photosynthesis and growth of some species while exerting no effect whatever on others. It thus seems more likely that pesticides and other toxic pollutants may influence the species composition of the phytoplankton than eliminate them entirely.

Much more significant than either of these considerations, however, is the implication in Cole's paper that the terrestrial ecosystem, including man, is dependent on the ocean for its oxygen supply. In the sea, as on land, the production and consumption of organic matter are in a steady state of equilibrium. Photosynthesis is balanced by respiration and oxygen production by utilization. The only exceptions are the yields of organic matter to man in the form of fish (subsequently oxidized on land) and the small net annual production of organic matter in the sea that results from its burial in the sediments. This is estimated to be about 0·1 per cent of the annual rate of photosynthesis or 5×10^7 tons (dry weight) buried each year with an equivalent net production of 15×10^7 tons of oxygen[4].

According to Hutchinson[5], "the oxygen content of the air is to be regarded as determined by a steady state between photolysis and deposition of reduced carbon, on the one hand, and fossilization, on the other", and "it is very unlikely, at least during phanerogoic time, that any very significant changes in the oxygen content of the atmosphere have taken place". The amount of oxygen in the atmosphere is placed at 12×10^{14} tons[6].

If, then, the net annual production of 15×10^7 tons of oxygen from the ocean is needed to maintain the steady state of oxygen in the atmosphere, and if all photosynthesis in the sea were to stop today, the total quantity of oxygen in the atmosphere could decrease by 10 per cent in no less than one million years. While I hasten to add that I do not advocate the elimination of life in the sea, I would submit that its effect on the Earth's oxygen supply would be less drastic than commonly held in today's atmosphere of eco-catastrophe.

This work was supported in part by the US Atomic Energy Commission.

[1] Cole, L. C., *Bioscience*, 18, 679 (1968).
[2] Whitaker, R. H., *Communities and Ecosystems* (Macmillan, New York, 1970).
[3] Menzel, D. W., Anderson, J., and Randtke, A., *Science*, 167, 1724 (1970).
[4] Emery, K. O., *Proc. Sixth World Petroleum Cong.*, Sec. 1, Paper 42, PD. 2, 483 (1963).
[5] Hutchinson, G. E., *The Solar System*, 2, *The Earth as a Planet*, 371 (Univ. Chicago Press, 1954).
[6] Rubey, W. W., *Bull. Geol. Soc. Amer.*, 62, 1111 (1951).

Interaction of pesticides with
aquatic microorganisms and plankton

By

GEORGE W. WARE and CLIFFORD C. ROAN

I. Introduction

The interactions of pesticides and soil microorganisms are heavily documented from the agricultural view. Many of the same soil particles, microorganisms, and pesticides are found in freshwater and estuarine ecosystems, and similar relationships may exist. It is the purpose of this paper, then, to review the interactions of pesticides and aquatic microorganisms, those microscopic plants and animals found in freshwater, estuarine, and marine environments.

a) Definition

Aquatic microorganisms and plankton, by the authors' definition, are comprised of those microflora and microfauna commonly found in fresh water, brackish, and marine environments. These include (1) the higher protists, or unicellular organisms, which are the algae, protozoa, fungi, and slime molds, and (2) the lower protists, blue-green algae, myxo-bacteria, spirochetes, and eubacteria. Plankton are organisms found floating or drifting almost passively and are carried about by wave action and currents. The phytoplankton are composed generally of algae to which the diatoms belong. Zooplankton are mostly very small and belong generally to the Protozoa, and animal phyla Arthropoda, Porifera (sponges), Cnidaria (coelenterates), Platyhelminthes (flatworms), Aschelminthes (roundworms), and Annelida (segmented worms). Plankton may also involve immature or larval stages of organisms normally excluded from this category, as well as microflora and fauna relocated from their original terrestrial surroundings.

Aquatic microorganisms are in many instances "weeds" in that they are merely soil microorganisms out of place. Fresh surface waters are usually not identified with any characteristic bacterial flora (FROBISHER 1949). The kinds of bacteria in them depend on their mineral and organic content, the soils with which they are in contact, surface pollution, and other factors. Many of the higher bacteria are common in fresh water lakes. Rainfall increases the numbers of soil bacteria in waters which collect runoff. Waters polluted with sewage have broad population ranges of *Escherichia coli* and other Enterobacteriaceae as well as Enterococci and *Clostridium perfringens*. Fecal pollution contributes such soil saprophytes as *Spirillium, Sarcina, Micrococcus, Mycobacterium, Bacillus*, yeasts, molds, Actinomycetes, and many others. The extent and nature of the bacterial populations in streams and rivers where sewage treatment plant effluents are found should differ significantly from mountain lakes and streams derived from melting snow.

b) Entry of pesticides into aquatic environments

WESTLAKE and GUNTHER (1966) classified environmental contamination under two categories, *i.e.*, (1) intentional (direct) application and (2) unintentional (indirect) contamination. In Table I (WESTLAKE and GUNTHER 1966), we have reclassified these sources for the aquatic environment. Our classification is perhaps more stringent (or the result of better hindsight) than the above-referenced authors in that many of the sources they classed as unintentional we have classified as intentional. This classification, we feel, is justified because pesticides were in fact knowingly added to the aquatic environment. What was unknown was their potential persistence and the as yet incompletely comprehended long range consequences of such additions.

For instance, the association of run-off pesticides with organic particulate matter in estuaries is significant because many organisms rely on these particles for part or all of their energy requirements. Accord-

Table I. *Sources of pesticides in the aquatic environment*

A. Intentional introductions
  1. Control of objectional flora and/or fauna
  2. Industrial wastes
    a. Pesticide manufacturers and formulators
    b. Food industry
    c. Moth-proofing industry
  3. Disposal of unused materials
  4. On-site field cleaning of application, mixing, and dipping equipment
  5. Disposal of commodities with excessive residues
  6. Decontamination procedures
B. Unintentional introductions
  1. Drift from pesticide applications to control objectional flora or fauna
  2. Secondary relocation from target area via natural wind and water erosion
  3. Irrigation soil water from target areas
  4. Accidents involving water-borne cargo
  5. Application accidents involving missed targets or improper chemicals

ing to Odum *et al* (1969) plant detritus consumers include amphipods, isopods, harpacticcid copepods, various filter- and deposit-feeding bivalves, annelid worms, caridean and penaeid shrimp, fiddler crabs, and mullets. DDT and its metabolites accumulate in plant detritus within estuaries and may persist there for many years. The residues appear to be most abundantly associated with particulates having diameters from 250 to 1,000 microns (μ).

A detailed recitation of the pesticides detected in the aquatic environment at various places and times would include all pesticides that have been or are in current use. The persistence of various pesticides in the nonliving components of the aquatic environment, while beyond the scope of the present review, must be recognized particularly where this persistence may be greater than in the soil. A specific example, from Schwartz (1967), of the metabolism of 2,4-D being much slower in aqueous than in soil environments is illustrative of this situation.

In addition to possible physical/chemical reactions of pesticides in water with the nonliving components two other interaction systems are of interest, *i.e.*, the direct action of the pesticide on the flora and fauna of the environment, and the effects of the flora and fauna on the pesticides.

II. Toxicity of pesticides to aquatic microorganisms

Toxicity is frequently regarded only as the lethal effects of the chemical upon a particular organism. Other direct effects that are included in this review concern changes in growth rate and changes in specific metabolic rates, *i.e.*, photosynthesis. Indirect actions that are ecologically important are the results of stress on one or more organisms that permit previously suppressed competitors to flourish, further stressing dominant populations.

The diversity of species in both the fresh water and marine environments is even greater than the diversity of chemicals added to these environments. This diversity of organisms has several important consequences giving rise first of all to a necessary multiplicity of experimental systems and techniques for evaluation of the toxicity of environmental contaminants. A further consequence of major importance is the tendency, particularly of the public (and only slightly less on the part of the scientist), to extrapolate from the adverse effects of certain chemicals or classes of chemicals on a few organisms to include an entire ecosystem.

The diversity of laboratory tests is represented by the designs employed by WURSTER (1968) where cultures of the test phytoplankton were exposed to different concentrations of DDT for 20 to 24 hours, with 14 hours of light and ten hours of dark. After this exposure period C^{14}-bicarbonate was added and the algae were illuminated for an additional four to five hours. Radio-assay of the filtered cultures was employed to estimate the amount of carbon fixed by photosynthesis. The effects of DDT exposures were determined by comparisons of the exposed and control populations.

In contrast, UKELES (1962) prepared the toxicants as concentrated or saturated stock solutions in sterile sea water. After at least 24 hours, serial dilutions in sterile sea water were prepared and appropriate amounts were added to the sterile basal medium. Cultures were incubated for ten days under continuous illumination. Growth rates were estimated by determination of transmittance at 530 mμ.

The excellent tabulation of LAWRENCE (1962) in the data pertaining to microflora and fauna and planktonic forms in general lists such diverse test environments as aquaria, distilled water, mud, and ponds, with exposures to the herbicides ranging from a few hours to 60 days. Effects are described as lethal, percent killed, not toxic, and tolerated.

The effects of a series of pesticides on estuarine phytoplankton were investigated by BUTLER (1965 a and b). Some of these results, together with data from CIRCULAR 167 (1963) appear in Table II. These data were based on a four-hour exposure to the toxicants in question. The organisms were described only as phytoplankton. Pesticides at concentrations of 1.0 part per million (p.p.m.) reducing carbon fixation by 75 percent or more are comprised of six herbicides, 11 insecticides (two organophosphorus and nine organochlorine), and two fungicides. Of the 26 herbicides in Table II, nine are listed as having no adverse effect under the test conditions. Of 25 insecticides and the three fungicides tested all had adverse effects on phytoplankton.

Further data on the effects of DDT on photosynthesis by four species of phytoplankton are presented in Table III (WURSTER 1968). These data suggest that a specific adverse effect from DDT concentrations as low as 0.1 p.p.m. may be expected. WURSTER (1968) suggests the absence of a threshold concentration of DDT on the basis of a

160

Table II. *Effects of pesticides at concentrations of one p.p.m. on estuarine phytoplankton*

Pesticides [a]	Decrease of carbon fixation during 4 hours exposure (%)
Herbicides [b]	
Monuron (urea)	94
Neburon (urea)	90
2,4,5-T, polyglycol butyl ether ester (phenoxy)	89
Diuron (urea)	87
Silvex (phenoxy)	78
DEF (defoliant)	75
Zytron (phosphate)	59
Paraquat	53
2,4-D, 2-ethylhexyl ester	49
Diquat	45
2,4-D, propyleneglycol butyl ether ester	44
Fenuron	41
Dacthal	37
Tillam	24
2,4-D, butoxy ethanol ester	16
N-Serve	15
Hydram	9
Dalapon Na salt	0
Kurosal SL (60 percent Silvex)	0
Tordon	0
2,4-D acid	0
2,4-D dimethylamine salt	0
2,4,5-T acid	0
MCP amine	0
Eptam	0
Vernam	0
Insecticides	
Kepone	95
Heptachlor	94
Chlordane	94
Toxaphene	91
Ronnel	89
Thiodan	87
Methyl Trithion	86
Dieldrin	85
Aldrin	85
Methoxychlor	81
DDT	77
Ethion	69
Dibrom	56
Di-Syston	55
Endrin	46
Mirex	42
Bayer 37344	39
ASP-51	30
Lindane	28
Carbaryl	17

Table II. (continued)

Pesticides [a]	Decrease of carbon fixation during 4 hours exposure (%)
Imidan	8
Demeton	7
Baytex	7
Malathion	7
Diazinon	7
Fungicides	
Ferbam	97
Dyrene	91
Phaltan	32

[a] Insecticides and some other materials from ANONYMOUS (1963).
[b] Herbicides from BUTLER (1965 a).

sigmoid dose-response curve in the case of *Skeletonema costatum* where the concentration of DDT was constant but the cell population density varied. This suggestion does not appear warranted in view of the presented data indicating the absence of any clearly definable effect at concentrations at or below 1.0 part per billion (p.p.b.).

UKELES (1962) presented data on the toxicity of 17 chemicals to five species of marine phytoplankton. Table IV is a rearrangement of data from this paper to indicate the highest concentrations preventing any growth of the cultures during a period of ten days. These data indicate that of the insecticides tested only toxaphene approaches the phytotoxicity of herbicides tested against marine phytoplankton.

PIERCE (1958 and 1960) found that Kuron applied to a fresh water pond had no apparent lasting effect on various plankton although there was a noticeable decrease in populations within 24 hours after treatment. The predominant forms observed were *Ceratium, Dinobryon, Synura,* and *Bosmina,* while *Spirogyra, Volvox, Daphnia, Tabellaria, Micrasterias,* and *Fragilaria* were frequently taken.

Table III. *Effects of DDT on photosynthesis by phytoplankton*

	C[14] uptake as % of controls	
	DDT concentration	
Organism	1.0 p.p.b.	100 p.p.b.
Skeletonema costatum	70	25
Coccolithus huxleyi	75	35
Pyramimonas sp.	80	20
Peridinium trochoideum	55	25

Table IV. *Concentrations of certain pesticides preventing growth in five species of marine phytoplankton*

Pesticide	Concentration in p.p.m. producing no growth				
	Proloccus sp.	*Chlorella* sp.	*Dunaliella euchlora*	*Phacodactylum tricornutum*	*Monochrysis lutheri*
Trichlorfon	1000	500	500	500	100
TEPP	500	500	500	500	500
Phenol	500	500	500	100	100
Dowacide A	100	100	100	100	50
Orthodichlorobenzene	13	13	13	13	13
Chloronitropane	80	80	80	80	80
PVP-iodine	>100	>100	50	50	100
Carbaryl	10	10	10	< 0.1	1
Nabam	10	10	1	1	1
Lindane	> 9	> 9	> 9	7.5	7.5
Toxaphene	.15	.07	0.15	.04	.01
DDT	> 60	> 60	> 60	> 60	60
Lignasan	.006	.006	.06	.06	.006
Fenuron	29	2.9	2.9	2.9	2.9
Neburon	0.2	0.2	0.2	0.2	<0.04
Monuron	0.02	0.02	0.02	0.02	0.02
Diuron	0.004	0.04	0.004	0.004	$<2 \times 10^{-5}$

COWELL (1965) in pond studies in New York observed that neither Silvex nor sodium arsenite was toxic or inhibitory to zooplankton at concentrations of two p.p.m. Zooplankton populations were drastically reduced by applications of nine p.p.m. of sodium arsenite. COPE (1966), however, in studies of a fresh-water ecosystem found sodium arsenite to have an LC_{50} of 1,800 p.p.m. for *Daphnia pulex*.

Selected data from COPE's (1966) investigations are presented in Table V. As might be expected the insecticides are much more toxic to *Daphnia* than are the herbicides. HARDY (1966) indicates that the herbicide Tordon at one p.p.m. was nontoxic to *Daphnia*.

TATUM and BLACKBURN (1962) indicate that diquat at 0.5 p.p.m. adversely affects plankton but that recovery is rapid. MULLIGAN (1967) lists copper sulfate, potassium permanganate, combinations of copper sulphate and silver nitrate, and simazine as nontoxic to zooplankton at concentrations that are toxic to phytoplankton. CABEJSZEK and STANISLAWSKA (1967) report Metasystox tested on organisms found in the Vistula River to be most toxic to *Daphnia* and that algae are the least sensitive. In investigations with methyl parathion in the same environment CABEJSZEK and STANISLAWSKA (1966) also found protozoa and daphnids to be more susceptible than filamentous algae.

Further evidence of the apparent toxicity of insecticides to aquatic plants as well as animals is found in LAZAROFF (1966) where algal development is inhibited by one p.p.m. of DDT or lindane as well as

Table V. *Relative toxicities of pesticides to Daphnia pulex*

Pesticide	Toxicity (48 hr. LC$_{50}$ in p.p.m.)
DDT	0.4
Diazinon	0.9
Malathion	2
Azinphosethyl	3
Carbaryl	6
Toxaphene	15
Endrin	20
Pyrethrins	25
Heptachlor	42
Dieldrin	250
Lindane	460
Trifluran	240
Diuron	1,400
Sodium arsenite	1,800
Silvex	2,400
2,4-D	3,200
Dichlobenil	3,700
Fenac	4,500

the fungicides captan or Nabam. He found that thiocarbamate pesticides interfere with photo-induced development in the blue-green algae *Nostoc muscorum* A. Although parathion at concentrations of ten p.p.b. inhibits motility in *Euglena gracilis*, it does not prevent growth.

Investigations of the effects of pesticides on bacteria in aquatic environments are not so numerous. LICHTENSTEIN *et al.* (1966) include data on the abundance of bacteria in water stored for three months containing various insecticides. Table VI (LICHTENSTEIN 1966) indicates that DDT apparently does not have an adverse action, as do the

Table VI. *Changes in bacterial populations after storing for three months in one p.p.m. aqueous solutions of pesticides* (LICHTENSTEIN 1966)

Pesticide	Bacteria/ml. in control / bacteria/ml. in treatment	
	Lake water	Soil water *
Parathion	0.43	0.62
Methyl parathion	1.14	0.62
Azinphosmethyl	0.25	0.47
Dieldrin	2.46	0.41
DDT	1.57	2.04
BHC	0.17	0.57

* Water percolated through soil before adding pesticides.

other compounds listed, on bacteria derived from soil, while in the case of lake water only parathion, Guthion, and BHC have adverse effects. LUCZAK and MALESZEWSKA (1967) report that Metasystox at seven p.p.m. does not significantly affect the growth rate of bacteria.

BUTLER and SPRINGER (1963) cite experiments with chronic exposures of three to six months (Table VII) in which test animals were maintained in running sea water to which the pesticide was continuously added. These authors state, "The unfiltered sea water supplying these aquaria contained planktonic larvae. Large numbers of these benthic animals, including at least 25 species in 7 phyla, set fortuitously and grew equally well in the experimental and control aquaria." JONES and MOYLE (1963) found that average counts of cladocerans, copepods, ostracods, rotifers, and *Volvox* were of the same general magnitude for treated and control ponds following field tests with DDT at one lb./acre for mosqutio control.

KASAHARA (1962) found that Dipterex at ten p.p.m. had no adverse effects on phytoplankton. Concentrations of 0.2 p.p.m. killed all crustacea except certain copepods while rotifers, flagellates, and ciliates were not affected.

Copper is an essential minor element for microorganisms and is commonly added in concentrations of 0.2 to 2.0 p.p.m. to synthetic culture media, and as a general rule, bacterial numbers increase with organic matter in sediments as well as in soils. However, residual copper from copper sulfate weed treatment in two Oregon coastal lakes effectively decreased the numbers of bacteria in bottom samples taken two years later, according to WATSON and BOLLEN (1952). This inhibition was probably caused by 11 to 30 p.p.m. of total copper and occurred in the presence of large amounts of organic material. Controls (four to seven p.p.m. of copper) gave much higher counts even when organic matter was low. The results are anomolous considering the generally low toxicity of copper for bacteria, plus the copper-binding characteristic of organic matter.

CRANCE (1963) stated that the reduction of blue-green algae in a

Table VII. *Concentrations of insecticides in running sea water*

Pesticide	Concentration (p.p.m.)	Test animal
DDT	1	Juvenile clam
Dieldrin	5	Juvenile clam
Dieldrin	0.1	Spot
Dieldrin	0.01	Spot
Dieldrin	0.001	Spot
Aldrin	2	Oysters and mussels
Malathion	2	Oysters and mussels
Toxaphene	50	Oysters and mussels

fresh water pond should be beneficial to fish, in that water bacteria usually increase enormously after algae decay. Protozoa then feed on the bacteria followed by an increase in Rotifera and Crustacea used by fish as food. The number of zooplankters usually increased within a few days after an application of copper sulfate (CRANCE 1963). Copepods were the predominant species, followed by cladocernas. Rotifers, Chaoboridae larvae, and ostrocods were the remaining zooplankters, and were usually more abundant after applications of copper sulfate.

Toxicity, when expressed inversely, is the flourishing of one group of microorganisms with the suppression of others. A distinct increase in the blue-green algae populations was observed when lindane was applied at five, six, and 50 kg./ha. to submerged Philippine rice soils (RAGHU and MACRAE 1967 a), which they attributed to elimination of small algae-eating animals. They also observed (1967 b) that lindane at six kg./ha. resulted in significant increases in nitrogen fixation, and higher populations of anaerobic, phosphate-dissolving bacteria.

Diazinon applied at two or 20 kg./ha. was found to significantly stimulate the actinomycete and the algal populations in submerged Philippine rice soils (SETHUNATHAN and MACRAE 1969). Since most investigations seek and anticipate population decreases there may be many unobserved flourishing species in such tests.

III. Concentration

a) Freshwater microorganisms

The water solubilities of most pesticides, particularly the organic insecticides, are generally quite low varying from about·one p.p.b. to complete miscibility with water. Their relative solubilities from least to greatest are the organochlorine, carbamate, and organophosphate insecticides (GUNTHER et al. 1968).

Pesticides, partly because of their very low water solubilities, tend to seek living organisms. Microscopic plants and higher aquatic florae quickly accumulate quantities of pesticides from the water medium and retain them in and on their tissues. Aquatic fauna of all sizes similarly tend to remove pesticides from the water and store them (COPE 1966).

In certain macroflorae submersed plant leaves accumulated relatively high amounts of the herbicides endothall or diquat from 0.1 p.p.m. concentrations. Disodium endothall was absorbed more rapidly and in larger amounts by leaves of susceptible plants, American and sago pondweed (*Potamogeton nodosus* and *P. pectinatus*) than by leaves of the resistant plant, American elodea (*Elodea canadensis*). Elodea leaves, however, absorbed more diquat than the sago (SEAMAN and THOMAS 1966).

In still smaller plants which begin to approach the definition of "aquatic microorganisms," WARE *et al.* (1968) observed that a filamentous alga and a pondweed (Cladophora and Potamogeton) were better indicators of surface irrigation water contamination than the alga Ocillatoria.

Bottom fauna in the deep water of the Great Lakes are dominated by the amphipod, *Pontoporeia affinis.* This species is an important food source for fish and long-tailed ducks in Lake Michigan. Pooled samples of this crustacean averaged 0.41 p.p.m. for DDT, DDE, and TDE, or 50 times the level found in surrounding muds. Fish which fed chiefly on the amphipods, namely alewives, chub, and whitefish, had about ten times, while the ducks had 15 times that of *Pontoporeia* (HICKEY *et al.* 1966).

Three species each of fungi, streptomycetes, and bacteria were tested for their ability to accumulate DDT and dieldrin from distilled water (CHACKO and LOCKWOOD 1967). After four hours the fungi accumulated 75 percent of the dieldrin and 60 to 83 percent of the DDT. One bacteria, *Agrobacterium tumefaciens,* accumulated 90 percent of the dieldrin and 100 percent of the DDT, while the results from the others were inconclusive, probably due to losses through cell washing. Using an autoclaved *Streptomycetes,* the authors concluded that heat-killed and living mycelia accumulated most of the insecticides from the medium, indicating that the accumulation was probably physical and did not involve metabolism. The accumulation of pesticides by microbes in soil and subsequently in water, is probably a strong factor in the retention of these compounds.

Cultures of a blue-green alga (*Anacyctis nidulans*), a green alga (*Scenedesmus obliquus*), a flagellate (*Euglena gracilis*), and two ciliates (*Paramecium bursaria* and *P. multimicronucleatum*) were exposed separately to DDT or to parathion at one p.p.m. (GREGORY *et al.* 1969). These algae and protozoa concentrated DDT 99 to 964 times and parathion 50 to 116 times during a seven-day exposure period. *P. multimicronucleatum* absorbed the highest levels of both insecticides. This indicated that an organism feeding on these unicellular forms may receive a higher level of insecticide than directly from water. Only small amounts of DDT or parathion remained in the supernatent liquid after seven days. No metabolities were detected of either compound in any of the organisms. The possibility of insecticide metabolism having occurred should not be excluded considering the one-step hexane extraction method employed.

In a similar study, using mycelial fragments and soil instead of a water medium, the accumulation and concentration of dieldrin, DDT, and the herbicide PCNB above ambient levels of mycelia of actinomycetes and fungi was demonstrated (Ko and LOCKWOOD 1968). However, in soil the total amount of these compounds accumulated by mycelia was relatively small, the highest being ten percent when a large amount of mycelium was used. This indicates that the ability to take up chlorinated pesticides may be a generally nonspecific property of cells of actinomycetes, bacteria, and fungi. Here the differences in the accumulation or absorption from their respective media are explained by the competitive adsorption by soil particles and a lack of continuing surface exposure to soil as in the aqueous medium.

Copper sulfate was absorbed within 72 hours and removed from solution by a heavy bloom of algae. These ponds in New Jersey were treated to yield 0.5 p.p.m. of copper but it will be noted in Table VIII

Table VIII. *Typical concentration of copper in New Jersey ponds treated with copper sulfate to yield 0.5 p.p.m. of copper* (TOTH and RIEMER 1968)

Water depth	Concentration (p.p.m.) after					
	Initial	2 hr.	4 hr.	24 hr.	48 hr.	72 hr.
Top	0.05	0.14	0.14	0.13	0.11	0.11
Bottom	0.03	0.06	0.06	0.13	0.13	0.09

(TOTH and RIEMER 1968) that at no time did the actual concentration approach the intended level.

b) Marine microorganisms

Pesticides applied directly to estuarine waters are absorbed by plankton almost immediately (BUTLER 1967). Here it was observed that the biological magnification of residues in the food web could progress from an estimated 1.0 p.p.b. of DDT and related metabolites in the water to 70 p.p.b. in plankton to 15 p.p.m. in fish and up·to 800 p.p.m. in porpoise blubber.

A marine diatome (*Cylindrotheca closterium*) absorbed and concentrated DDT up to 190-fold from its medium containing 0.1 p.p.m. of DDT (KEIL and PRIESTER 1969).

DDT residues in the soil of an extensive Long Island salt marsh averaged more than 13 lb./acre, with a maximum of 32 lb. A systematic sampling of various organisms showed concentrations of DDT increasing with trophic level through more than three orders of magnitude. When the bottom mud held 0.28 p.p.m., zooplankton contained 0.04 p.p.m. and the ringbilled gull 75 p.p.m. (WOODWELL et al. 1967).

Measuring dieldrin and DDE in micro- and macrozooplankton off the Northumberland Coast, ROBINSON et al. (1967) concluded that there appeared to be a close correlation with residues and their trophic level. The lowest residues were found in the second while the highest were recovered from animals in the fifth. From their survey they determined that marine animals did not appear to be better indicators of environmental contamination by insecticides than terrestrial animals.

IV. Metabolism of Pesticides

Because of the previously established relationship of many soil and aquatic microorganisms, the discussion must again be somewhat all-inclusive. That is, if pesticide metabolism occurs with a particular organism in its terrestrial environment quite naturally it should occur similarly qualitatively, though not necessarily quantitatively, in its aquatic surroundings. The essential differences appear to be usually fewer nutrients/unit mass in water than soil, thus probably less biochemical activity, and the much more rapid food chain turnover in the aquatic than in the terrestrial environment.

Many pesticides are resistant to microbial action, and either they remain unaltered, even in the presence of a large and active microbial population, or they are metabolized at a disturbingly slow rate (Table IX) (ALEXANDER 1964).

Table IX. *Persistence of insecticides in some soils* (ALEXANDER 1964)

3 years	5 years	11 years
DDT	Parathion	BHC
Dieldrin	Lead arsenate	Chlordane
Toxaphene		

Several good reviews have been compiled relating to the interaction of soil microorganisms and pesticides (MARTIN 1966, ALEXANDER 1964, EDWARDS 1966). Microorganisms in soils decompose or metabolize many of the herbicides (AUDUS 1964, FUNDERBURK and BOZARTH 1967, THEIGS 1962, MACRAE and ALEXANDER 1965), fungicides (MUNNECKE 1966), organophosphate insecticides (AHMED and CASIDA 1958, GUNNER and ZUKERMAN 1968, ROBERTS et al. 1962), organochlorine insecticides (CHACKO et al. 1966 and 1967, EDWARDS 1966), and antibiotics (GOTTLIEB 1952, PRAMER 1958).

Many species of protists have been identified which metabolize herbicides and insecticides (AUDUS 1964, MUNNECKE 1966, THIEGS 1962, ALEXANDER 1964). A list of genera represented in these reviews is shown in Table X.

a) Organochlorine insecticides

Yeasts are universal in distribution and are known to metabolize DDT (KALLMAN and ANDREWS 1963). They observed that C^{14}-DDT was converted only to DDD (TDE) by reductive dechlorination. No DDE was observed, and when DDE was incubated with yeast cultures only DDE was recovered.

The marine diatome, *Cylindrotheca closterium*, converted DDT only to DDE (nine percent) after three weeks in a medium containing

Table X. *Genera of selected microorganisms known to metabolize pesticides*

Bacteria	Actinomycetes	Fungi	Algae	Yeasts
Achromobacter	Mycoplana	Acrostalagmus	Chlamydomonas	Saccharo-
Aerobacter	Nocardia	Aspergillus	Chlorella	myces
Agrobacterium	Streptomyces	Clonostachys	Cladophora	Torulopsis
Alcaligenes		Cylindrocarpon	Cylindrotheca	
Arthrobacter		Fusarium	Oscillatoria	
Azotobacter		Geotrichum	Vaucheria	
Bacillus		Glomerella		
Clostridium		Helminthosporium		
Corynebacterium		Mucor		
Escherichia		Myrothecium		
Flavobacterium		Penicillium		
Klebsiella		Stachybotrys		
Micrococcus		Trichoderma		
Paracolobactrum		Xylaria		
Proteus				
Pseudomonas				
Rhizobium				
Sarcina				
Serratia				
Sporocytophaga				
Thiobacillus				
Xanthomonas				

0.1 p.p.m. of DDT. No other metabolites were found (KEIL and PRIESTER 1969).

Two coliform bacteria (*Escherichia coli* and *Aerobacter aerogenes*), commonly found in aqueous environments, were found to metabolize DDT to TDE (MENDEL and WALTON 1966). In this particular study these cultures were isolated from rat feces, and resulted in the indication that the normal florae of the gastrointestinal tract are the major source of TDE found in animals fed DDT, rather than the liver.

In a similar study using the above two plus a third facultatively anaerobic species (*Klebsiella pneumonia*), WEDEMEYER (1966) indicated *A. aerogenes* as the most effective in the reductive dechlorination of DDT to TDE. A detailed study revealed that reduced $Fe(II)$ cytochrome oxidase was responsible for DDT dechlorination. WEDEMEYER proposed that this may possibly explain the persistence of DDT in aerobic soils.

Anaerobic conversion of DDT to TDE in soil was reported by GUENZI and BEARD (1967). DDT was added to soil and the mixture was incubated anaerobically for two and four weeks. DDT and seven possible metabolic products were separated by thin-layer chromatography. The DDT was dechlorinated by unidentified soil microorganisms to TDE and only traces of other degradation products were detected (Table XI) (GUENZI and BEARD 1967). No degradation of DDT was detected in sterile soil.

Table XI. *Degradation products recovered from soil treated with DDT and incubated anaerobically* (Guenzi and Beard 1967)

	Recovery (μg.) after	
Materials found	2 weeks	4 weeks
DDA	0.37	0.51
p-Chlorobenzoic acid	0.24	0.59
Dicofol	0.15	0.61
4,4'-Dichlorobenzophenone	0.39	0.64
TDE	7.1	35.0
DDT	62.0	19.0
4,4'-Dichlorodiphenylmethane	0.09	0.03
DDE	0.19	0.25
Total	71.0	57.0

HILL and McCARTY (1967) conducted an extensive study of organochlorine insecticide degradation under anaerobic conditions. The pesticides were mixed continuously at 35°C. with biologically active anaerobic digested wastewater sludge obtained from a sewage treatment plant. The anaerobic conditions were ideal in that an active culture of anaerobic, methane-producing and sulfate-reducing bacteria were present. The metabolism of most of the organochlorine insecticides was more rapid under anaerobic than under the corresponding aerobic conditions, with the exception of heptachlor epoxide and dieldrin, which were very stable in both conditions. Extractable metabolic products were more common in the anaerobic than in the corresponding aerobic conditions. Under anaerobic conditions the order of increasing persistence or stability of the chlorinated insecticides was lindane, heptachlor, endrin, DDT, TDE, aldrin, heptachlor epoxide, and dieldrin. DDT was metabolized very rapidly to TDE under anaerobic conditions, but persisted as DDT under aerobic conditions of several mg./l. of dissolved oxygen. An increase from 20° to 35°C. produced no significant increases in degradation rates except for the anaerobic metabolism of lindane and the aerobic metabolism of DDT. These insecticides were also more strongly sorbed by algae than by bentonite clay. The implications of this study are that DDT would be readily converted to TDE on lake bottoms due to rapid anaerobic conversion by ooze organisms. Also implied is that lindane, aldrin, DDT, TDE, heptachlor, and possibly endrin would undergo a significant amount of degradation in anaerobic muds of algae pounds or in natural waterways which reach temperatures of 20 C. or warmer.

In the above study the term "degradation" was used in a broad sense to refer to any measurable chemical change in a pesticide under natural environmental conditions. Metabolites or degradation products other than DDT to TDE were not identified. The algae used were a mixed culture consisting primarily of *Vaucheria sessilis*, a filamentous species. The implication here is that algae eventually die and frequently settle to the bottom where they may decay, along with chlorinated insecticides, anaerobically.

Water from Clear Lake, California, bovine rumen fluid, and porphyrins under anaerobic conditions were all found to convert DDT to TDE (MISKUS et al. 1965). Lake water containing large amounts of plankton converted more DDT to TDE than did small amounts. The authors believed that variation in oxygen content of water in different areas of the lake and different rates of oxygen depletion on incubation, caused by different population levels of florae and fauna, may have contributed to the wide range of TDE levels found. Distilled water showed no conversion of DDT to TDE, true also in boiled rumen fluid. Porphyrins under the proper, anaerobic, reducing conditions can convert DDT to TDE, which the authors indicated as one possible mechanism to explain this conversion in many biological systems.

In culture solutions, most of the actinomycetes and filamentous fungi tested degraded PCNB; several actinomycetes dechlorinated DDT to TDE, but no microorganisms degraded dieldrin (Table XII)

Table XII. *Degradation of DDT, dieldrin, and PCNB by actinomycetes and fungi in culture solutions*

Microorganisms	Extent of degradation [*]		
	DDT	Dieldrin	PCNB
Fungi			
Aspergillus niger	—	—	+
Fusarium solani f. phaseoli	—	—	+
Glomerella cingulata	—	—	+
Helminthosporium victoriae	—	—	+
Mucor ramannianus	—	—	+
Myrothecium verrucaria	—	—	+
Penicillium frequentans	—	—	+
Trichoderma viride	—	—	+
Actinomycetes			
Nocardia sp.	++	—	+
Streptomyces albus	—	—	—
S. antibioticus	+	—	++
S. aureofaciens	++	—	+++
S. cinnamoneus	++	—	+
S. griseus	—	—	+
S. lavendulae	—	—	++
S. venezuelae	—	—	+
S. viridochromogenes	++	—	+

[*] = no. detectable degradation, + = less than ten percent, ++ = ten to 25 percent, and +++ = 25 to 50 percent.

172

(CHACKO et al. 1966). Streptomyces aureofaciens degraded PCNB to pentachloroaniline. Degradation of the pesticides in culture occurred only during the active growth phase of the actinomycetes or fungi, and stopped completely when growth ceased.

Dieldrin has since been shown to break down under the influence of a soil organism (MATSUMURA et al. 1968). This insecticide, which is highly stable in the presence of most microorganisms, was shown to metabolize in the presence of Pseudomonas sp. (Shell 33) originally isolated from a soil sample from the dieldrin factory yards of the Shell Chemical Company near Denver, Colorado. There were five major metabolites, with three lesser ones isolated. Surprisingly, one of the metabolites was identified as aldrin, the precursor to dieldrin, preceding its epoxidation in aerobic soils.

Lindane was actively degraded by unidentified microflora in flooded Philippine rice soils, but not when the soils were sterilized (RAGHU and MacRAE 1966). A second application of lindane 55 days later disappeared more rapidly than the first. The alpha, beta, gamma, and delta isomers disappeared from these flooded soils in 70 to 90 days MacRAE et al. 1967), and were believed to be degraded more actively by anaerobic than aerobic soil microflora under these conditions.

Two algae (Chlorella vulgaris and Chlamydomonas reinhardtii) metabolize lindane by dehydrochlorination to pentachlorocyclohexene, a non toxic lindane metabolite (SWEENEY 1968).

MacRAE et al. (1969) have observed the rapid anaerobic degradation of lindane by Clostridium sp. to a short-lived metabolite detectable by electron capture gas chromatography, definitely identified as not being pentachlorocyclohexene. After 27 hours incubation 75 percent of the theoretical amount of chlorine in the lindane was released by the bacteria as chloride ion in the reaction mixture.

b) Organophosphate insecticides

Four soil microorganisms, frequently identified in fresh water, were grown in pure cultures under aseptic conditions and exposed to organophosphate insecticides in metabolism studies by AHMED and CASIDA (1958). They were a yeast (Torulopsis utilis), two bacteria (Pseudomonas fluorescens and Thiobacillus thiooxidans), and a green alga (Chlorella pyrenoidosa). Rate studies were made on hydrolysis and oxidation of organophosphates added to the cultures as emulsions at 1,000 p.p.m. The compounds used were parathion, ronnel, dimefox, schradan, phorate, American Cyanamid 12008 (the propylthio analog of phorate) and the sulfinyl and sulfonyl derivatives, and the phosphorothiolate forms of phorate and AC 12008.

Chlorella and Torulopsis reacted similarly with the phosphorothioates, in that these toxicants were rapidly absorbed by the organisms and slowly released from living and dead cells in culture. With both, the S-(alkylthio)-methyl derivatives were more rapidly hydrolyzed than the S-(alkylsulfonyl)-methyl, which was slightly faster than the S-(alkylsulfinyl)methyls. In every instance the phosphorodithioate sulfoxides were the most and the phosphorothiolate sulfides the least stable to metabolism by these microorganisms.

Torulopsis and *Chlorella* oxidized the sulfides to sulfoxides, but *Chlorella* more effectively oxidized the dithioates to the thiolates. Little oxidation occurred with parathion, ronnel, dimefox, and schradan exposed to *Chlorella*.

The two bacteria did not oxidize, but did effectively hydrolyze phorate. The *Thiobacillus* did not utilize the sulfur in phorate.

Inquiring into the reasons for varying mosquito larvicidal longevity in the field, YASUNO *et al.* (1965) examined the effects of selected microorganisms on both organochlorine and organophosphate insecticides. Fenitrothion (Sumithion) used as mosquito larvicide was effective at 0.01 p.p.m., but the activity was lost after a few days. Ronnel, diazinon, fenthion, dieldrin, lindane, and DDT held up nicely for eight days, while fenitrothion, parathion, and methyl parathion were rapidly deactivated in bacteria-polluted water.

Among the organisms isolated, *Bacillus subtilis* was highly active in degrading fenitrothion, parathion, methyl parathion, and ronnel. Malathion, dichlorvos, and diazinon were not as readily degraded. Bacteria-free filtrates and sterilized cultures were inactive, indicating the living bacteria as the agent.

The main metabolic product was obtained by reduction of the nitro to the amino group. Amino-derivatives of all three compounds were identified after exposure to *B. subtilis*. The pH normally found in mosquito breeding places did not directly affect the insecticides.

Aspergillus oryzae was partly effective, but *Proteus vulgaris, Penicillium islandicum, Saccharomyces cereviciae*, and *Candida albicans* showed no effect in degrading fenitrothion.

Of 16 species of bacteria isolated from mosquito breeding waters, most degraded methyl parathion and fenitrothion into nontoxic products in one percent peptone culture solutions (HIRAKOSO *et al.* 1968). The insecticides studied were methyl parathion, fenitrothion, fenthion, diazinon, and dichlorvos (Table XIII).

Synergistic degradation of C^{14} ring-labeled diazinon by two bacterial genera was reported by GUNNER and ZUKERMAN (1968). *Arthrobacter* sp., known to attack the side chain of diazinon, was unable completely to metabolize the ring portion of the molecule. Similar tests with *Streptomyces* also indicated by it did not metabolize the pyrimidinyl carbon to carbon dioxide. However, when both were incubated together up to 20 percent of the C^{14} was captured as $C^{14}O_2$, suggesting the synergistic relationship between these organisms in metabolizing the pyrimidinyl moiety of diazinon. The only products identified in this study were carbon dioxide and the pyrimidinyl and phosphoro-

Table XIII. *Bacteria isolated from mosquito breeding waters and respective organophosphate insecticides degraded (HIRAKOSO et al. 1968)*

Bacteria	Deactivated insecticides
Pseudomonas aeruginosa	Fenthion
Serratia plymuthica	Dichlorvos
S. plymuthica	Methyl parathion
Escherichia coli	Fenitrothion
Paracolobactrum aerogenoides	Fenitrothion
Escherichia aurescens	Fenitrothion
Achromobacter eurydice	Fenitrothion
Pseudomonas convexa	Methyl parathion
P. putide	Methyl parathion
Escherichia freundii	Methyl parathion, fenitrothion, dichlorvos
Pseudomonas fluorescens	Methyl parathion
Serratia kiliensis	Methyl parathion
Flavobacterium aquatile	Methyl parathion

thioate moieties. The significant aspect of this study is the apparent synergistic relationship between two microorganisms in the metabolism of an insecticide molecule which neither can achieve alone.

Chlorella pyrenoidosa proteose, an alga found in aquatic environments as well as in the soil, has been shown as the responsible agent for metabolizing parathion around the roots of bean plants which in turn translocated the sulfur-containing metabolite to the above-ground parts (MACKIEWICZ et al. 1969). The algal metabolites consisted of 66 percent aminoparathion and a second product bearing the P—S group and the benzene ring. A trace amount of *p*-nitrophenol was found in the extracts but no paraoxon or *p*-aminophenol.

c) Herbicides

Metabolism of the five principal groups of herbicides has been discussed by KEARNEY (1966), classified into the phenylureas (fenuron, monuron, diuron), phenylcarbamates (CIPC, CEPC), *s*-trizaines (simazine, atrazine), chlorinated aliphatic acids (dalapon, TCA), and phenoxyalkanoic acids (2,4-D, 2,4-DB, 2,4,5-T).

The phenylureas will support the growth of bacteria *Xanthomonas, Sarcina, Bacillus,* and *Pseudomonas* and the fungi *Penicillium* and *Aspergillus.* In the metabolism of the phenylureas the dealkylation of methyl groups precedes hydrolysis of the urea linkage. To achieve this, first one methyl then the other is removed, followed by hydrolysis of urea to aniline. The resulting metabolic products are aniline, carbon dioxide, and ammonia.

The phenylcarbamates are generally less persistent than phenylureas. CIPC is degraded readily by *Pseudomonas striata, Flavobacterium* sp., *Agrobacterium* sp., and *Achromobacter* sp. CEPC is de-

175

composed by *Achromobacter* sp., and *Arthrobacter* sp. The enzymatic hydrolysis of CIPC by cell-free extracts of *P. striata* yields Cl-aniline, carbon dioxide, and isopropanol.

The *s*-triazines can be degraded by the fungus *Fusarium roseum,* especially atrazine. *Aspergillus fumigatus,* also a fungus, liberated $C^{14}O_2$ only from the side chain, but not from the ring. Side chain degradation occurs in culture solutions with the fungi *Fusarium roseum, Geotrichum* sp., *Trichoderma* sp., and *Penicillium* sp.

With the chlorinated aliphatic acids, soil persistence studies indicate that dalapon is rapidly degraded while TCA degrades more slowly. Soil microorganisms can dehalogenate and use the carbon for energy. At this point there were seven bacteria, five fungi, and two actinomycetes effective in decomposing dalapon. *Arthrobacter* sp. completely breaks down dalapon. An enzyme from these broken cells liberates the chlorine ion yielding pyruvic acid. TCA is dehalogenated by soil microorganisms, yet most isolates grow feebly on TCA as a sole source of carbon.

A great amount of work has been done on the phenoxyalkanoic acids or 2,4-D group. At least ten different organisms are reported to decompose 2,4-D. In these studies two pathways are described: (1) *beta*-oxidation of the alkanoic acid and (2) initial hydrolysis of the ether linkage between the ring and the side chain. Step (1) proceeds by sequential removal of two carbon fragments from the functional end of the alkanoic acid. The fate of the ring structure in soils has also been studied. Detection of 2,4-dichlorophenol, 4-chlorocatechol, and chloromuconic acid from either soil or pure culture studies suggests a sequence of reactions involving ring hydroxylation and cleavage and further metabolism of the open structure to carbon dioxide.

With the phenylcarbamates IPC, CIPC, and analogs, KAUFMAN (1966) reported on studies conducted in soil perfusion systems with mixed populations of soil microorganisms. He concluded that position, type, and number of halogen substituents are important factors affecting the microbial decomposition of both aliphatic and certain aromatic herbicides. The number of carbon atoms composing the aliphatic acid also is an important factor. The number of carbon atoms present in alkyl substituents of alkylphenylcarbamate pesticides affects their rate of decomposition. Degradation of *meta*-substituted chlorophenylcarbamates was more rapid than *ortho*-chloro-, *para*-chloro-, or dichloro-substituted phenylcarbamates.

C^{14}-labeled CIPC and 2,4-D were studied during their exposure to microbial degradation in aqueous solutions by Schwartz (1967). Under the conditions normally found in water supplies, these two materials were strongly resistant to chemical attack, and very little of each would adsorb onto suspended mineral solids. The microorganisms used here were obtained from the activated sludge unit and the effluent from the primary sedimentation basin of a water reclamation plant, though the organism counts or identities were not revealed. CIPC was definitely degraded by biological action. The addition of nutrient broth to the microbial systems increased greatly the rate at which the isopropyl-carbon atoms were metabolized, but had no effect on the rate of degradation of the phenyl group. SCHWARTZ (1967) proposed a metabolic pathway for the microbial degradation of CIPC by hydrolysis to 3-chloroaniline and isopropanol. The latter is completely metabolized. The 3-chloroaniline is modified to 4-chlorocatechol which is further degraded by way of the muconic acid-ketoadipic acid pathway typical for aromatic compounds.

The microbial breakdown of 2,4-D proceeded at a slow rate with only a small relative concentration degraded by a mixed microbial population in a dilute medium of salts. In solutions containing 0.1 and 1.0 mg./l. of 2,4-D, no more than 37 percent of the acetic acid moiety disappeared over a period of three to six months. The presence of large amounts of nutrient broth had no appreciable effect on the rate of decomposition.

On this basis the author concluded that 2,4-D would not be degraded materially by the microorganisms in a natural water supply. CIPC was considerably less resistant to biodegradation than 2,4-D, but may persist for weeks or months in a natural water environment. CIPC and 2,4-D have relatively short soil lives, but metabolism in water environments would be much slower than in soils. One might expect that the longer soil-residual chlorinated insecticides, such as dieldrin, endrin, DDT, and BHC, which remain in soil for several years, once reaching a water supply may persist for extremely long periods of time in the absence of microorganisms.

DeMarco et al. (1967) studied the behavior of 2,4-D at 50 p.p.b. in simulated stratified impoundments, using a natural river water. This herbicide was biologically degraded under warm-aerobic (22° to 26° C., and two mg. oxygen/l.), cold-aerobic (10° to 13°C, and two mg. oxygen/l.), and cold deoxygenated or anaerobic (10° to 13°C, <0.3 mg. oxygen/l.). The low temperature, as might be expected, reduced the rate of biodegradation. Low oxygen concentration reduced the degradation rate of 2,4-D more than temperature, while the combination of low temperature and low dissolved oxygen resulted in the least alteration, requiring 55 to 80 days to disappear. The warm-aerobic conditions accomplished the same task in six days, or a nine-fold increase in degradation rate. Total bacterial counts were made throughout the test, though not presented in the data or identified. These results indicate that certain environments present in impoundments could inhibit the degradation of certain pesticides.

V. Discussion

Contrasts in interpretation of toxicity data are apparent when comparisons are made of the data in Table V on the acute toxicity of DDT for *Daphnia pulex* (48 hr. LC$_{50}$ of 0.4 p.p.m.) and the comments cited

177

earlier from BUTLER and SPRINGER (1963) regarding the lack of adverse effect on growth of marine planktonic larvae in experiments involving three to six months exposures to DDT concentrations of one p.p.m. The reviewers recognize that identical species and test conditions were not involved; however, the situation reported by BUTLER and SPRINGER (1963) affords a more realistic basis for appraisal than data from acute toxicity tests.

Further basis for a conservative evaluation of acute data are found by contrasting the data concerning DDT (one p.p.m.) toxicity to estuarine phytoplankton, Table II, and those in Table III. The data in Table II indicate a decrease of 77 percent in carbon fixation while data in Table III for DDT at 0.1 p.p.m. indicate a mean decrease of 74 percent in C^{14} uptake. From these two incomparable values it is easy to conclude that a ten-fold increase in concentration produces very little increase in toxicity.

The complexity of ecosystems and the near simultaneous variety of man-induced environmental insults presents the scientist with a most difficult situation for evaluation. The complexities of the value judgments for the general public when confronted with diverse and often conflicting scientific conclusions are overwhelming.

The responses of microorganisms to and their reactions with pesticide components of their environment are a function not only of gross environmental dose and biochemical specificities of the microorganism but are also involved with size, shape, and gross composition of these life forms.

Surface area apparently plays an important role in the absorption of pesticides from water by microorganisms. From Table XIV (LAMANNA and MALLETTE 1965) it can be seen that one g. of yeast contains 8.3 x 10⁹ particles and has 9,100 sq. cm. of surface area. Greater still is the enterobacterium, *Escherichia coli*, which has 1.8 x 10¹² particles/g. yielding 56,000 sq. cm. of surface. The very high surface:mass ratio for microorganisms, compared to other aquatic or terrestrial plants and animals, in part explains their characteristic rapid and somewhat thorough absorption of pesticides from the aquatic environment.

Another point to be considered is that most pesticides, particularly

Table XIV. *Comparison of the approximate dimensions of some microorganisms* (LAMANNA and MALLETTE 1965)

Microorganism	Radius (μ)	Volume (μ^3)	Surface (μ^2)	Surface / volume	No. particles/g.	Surface area/g. (cm.²)
Saccharomyces cereviciae	3	110	110	1	8.3 x 10⁹	9.1 x 10³
Escherichia coli	0.5	0.52	3.1	6	1.8 x 10¹²	5.6 x 10⁴
E. coli phage	0.004	2.5 x 10⁻⁴	0.02	80	3 x 10¹⁵	6 x 10⁵

insecticides, are lipophylic, resulting in a selective partitioning into or onto a large surface area containing surface lipids. The ether- or chloroform-extractable lipids in microorganisms are shown in Table XV (PORTER 1946) indicating a range from 1.48 percent to 22.9 percent, depending on the organism and medium. These small forms contain all classes of lipids and lipoproteins, which are the main constituents of cell membranes and of certain intracellular structures such as mitochondria and chloroplasts.

Because of the recent appearance of the polychlorobiphenyls (PCB's) in the analytical literature as contaminants which may be mistakenly identified as some insecticide residues, it is not reasonable to assume that all residues of DDT and related metabolites reported in the foregoing papers are in fact as stated. The PCB's are well recognized for their false identities as DDT and other organochlorine insecticides and their derivatives by electron capture gas chromatography (LICHTENSTEIN et al. 1969, RISEBROUGH et al. 1968, REICHEL et al. 1969, RISEBROUGH 1969, REYNOLDS 1969. SCHECHTER (1969) is of the opinion that the PCB's are primarily a problem in analyzing samples from aquatic rather than terrestrial environments. COON (1969) reported finding the PCB's more in marine birds than in other samples. Thus it would appear to the reviewers that the residues of DDT reported from marine and estuarine conditions deserve cautious acceptance because of their potential mistaken identity with the PCB's

Table XV. *Lipid content of various microorganisms* (PORTER 1946)

Microorganism	% Lipid (dry wt.)
Ether extractable [a]	
Corynebacterium	4.9
Bacillus	4.4
Escherichia coli	3.6–7.9
Yeast	5.0
Oospora lactis	7.5–22.5
Aspergillus	2.6–13.0
Penicillium	4.13–22.9
Mucor	7.03
Chloroform extractable	
Bacillus	1.48
Escherichia	11.77
Klebsiella	7.36
Proteus vulgaris	7.10
Pseudomonas	10.67
Yeast	2.92

[a] Ether, petroleum ether, or alcohol-ether extractable.

Pesticide	Chemical designation
Aldrin	1,4:5,8-Dimethanonaphthalene, 1,2,3,4,10,10-hexachloro-1,4,4a,5,8,8a-hexahydro-, *endo-exo* isomer
ASP-51	tetra-*n*-propyl dithiopyrophosphate
Atrazine	*s*-Triazine, 2-chloro-4-(ethylamino)-6-(isopropylamino)-
Azinphosmethyl	Phosphorodithioic acid, *O,O*-dimethyl ester, S-ester with 3-(mercaptomethyl)-1,2,3-benzotriazine-4(3*H*)-one
Bayer 37344	Carbamic acid, methyl-, 4-(methylthio)-3,5-xylyl ester
Baytex	Phosphorothioic acid, *O,O*-dimethyl *O*-[4-(methylthio)-*m*-tolyl] ester
BHC	Cyclohexane, 1,2,3,4,5,6-hexachloro-
Carbaryl	Carbamic acid, methyl-, 1-naphthyl ester
CEPC	Carbamic acid, *N*-(3-chlorophenyl)-1-chloro-2-propyl
Chlordane	4:7-Methanoindan, 1,2,4,5,6,7,8,8a-octachloro-3a,4,7,7a-tetrahydro-, *endo*-isomer
CIPC	Carbanilic acid, *m*-chloro-, isopropyl ester
2,4-D	Acetic acid, 2-4-dichlorophenoxy-
Dacthal	Terephthalic acid, tetrachloro, dimethyl ester
Dalapon	Propionic acid, 2,2-dichloro-
2,4-DB	Butyric acid, 4-(2,4-dichlorophenoxy)-
DDA	Acetic acid, bis(*p*-chlorophenyl)-
DDE	Ethylene, 1,1-dichloro-2,2-bis(*p*-chlorophenyl)-
DDT	Ethane, 1,1,1-trichloro-2,2-bis(*p*-chlorophenyl)-
DEF	Butyl phosphorotrithioate
Demeton	Phosphorothioic acid, *O,O*-diethyl-*O*-[2-(ethylthio)ethyl] ester mixed with *O,O*-diethyl-*S*-[2-(ethylthio)ethyl] ester
Diazinon	Phosphorothioic acid, *O,O*-diethyl *O*-(2-isopropyl-6-methyl-4-pyrimidinyl) ester
Dibrom	Phosphoric acid, 1,2-dibromo-2,2-dichloroethyl-, dimethyl ester
Dichlobenil	Benzonitrile, 2,6-dichloro-
Dichlorvos	Phosphoric acid, 2,2-dichlorovinyl-, dimethyl ester
Dicofol	Benzhydrol, 4,4′-dichloro-alpha-(trichloromethyl)-
Dieldrin	1,4:5,8-Dimethanonaphthalene, 1,2,3,4,10,10-hexachloro-6,7-epoxy-1,4,4a,5,6,7,8,8a-octahydro-, *endo-exo* isomer
Dimefox	Phosphorodiamidic fluoride, tetramethyl
Dipterex	See Trichlorfon
Diquat	Dipyrido[1,2-*a*:2′,1′-*c*]pyrazidiinium compounds, "6,7-dihydro____. . ." dibromide
Di-Syston	Phosphorodithioic acid, *O,O*-diethyl *S*-[2-ethylthio)ethyl] ester
Diuron	Urea, 3-(3,4-dichlorophenyl)-1,1-dimethyl-
Dyrene	*s*-Triazine, 2,4-dichloro-6-(*o*-chloroanilino)-
Endothall	7-Oxabicyclo(2.2.1)heptane-2,3-dicarboxylic acid, disodium salt
Endrin	1,4:5,8-Dimethanonaphthalene, 1,2,3,4,10,10-hexachloro-6,7-epoxy-1,4,4a,5,6,7,8,8a-octahydro-, *endo-endo* isomer
Eptam	Carbamic acid, dipropylthio, S-ethyl ester
Ethion	Ethyl methylene phosphorodithioate
Fenac	Acetic acid, 2,3,6-trichlorophenyl-
Fenitrothion	Phosphorothioic acid, *O,O*-dimethyl *O*-4-nitro-*m*-tolyl ester

Table XVI. (continued)

Pesticide	Chemical designation
Fenthion	Phosphorothioic acid, O,O-dimethyl O-[4-(methythio)-m-tolyl] ester
Fenuron	Urea, 1,1-dimethyl-3-phenyl
Ferbam	Iron, tris(dimethyldithiocarbamato)-
Heptachlor	4:7-Methanoindene, 1,4,5,6,7,8,8a-heptachloro-3a,4,7,7a-tetrahydro-, endo-isomer
Heptachlor epoxide	4:7-Methanoindan, 1,4,5,6,7,8,8a-heptachloro-2,3-epoxy-3a,4,7,7a,tetrahydro-, endo-isomer
Hydram (Molinate)	S-ethyl hexahydro-1H-azepine-1-carbothioate
Imidan	Phosphorodithioic acid, O,O-dimethyl ester, S-ester with N-(mercaptomethyl)phthalimide
IPC	Carbamic acid, isopropyl-, N-phenyl ester
Kepone	1,3,4-Metheno-2H-cyclobuta[cd]pentalen-2-one, decachlorooctahydro-
Lignasan	ethylmercury phosphate
Lindane	Cyclohexane, 1,2,3,4,5,6-hexachloro-, gamma-isomer
Malathion	Succinic acid, mercapto-diethyl ester, S-ester with O,O-dimethyl phosphoro-dithioate
MCP	Acetic acid, 4-chloro-o-tolyloxy-
Metasystox	Phosphorothioic acid, O-[2-(ethylthio)-ethyl] O,O-dimethyl ester, mixed with S-[2-(ethylthio)ethyl] O,O-dimethyl ester
Methoxychlor	Ethane, 1,1,1-trichloro-2,2-bis(p-methoxyphenyl)-
Methyl parathion	Phosphorothioic acid, O,O-dimethyl O-p-nitrophenyl ester
Methyl Trithion	Phosphorodithioic acid, S-[(p-chlorophenyl)thiomethyl]-, O,O-dimethyl ester
Mirex	1,3,4-Metheno-2H-cyclobuta[cd]pentalene, dodecachlorooctahydro-
Monuron	Urea, 1,1-dimethyl-3-(p-chlorophenyl)-
Nabam	Carbamic acid, ethylene bisdithio-, disodium salt
Neburon	Urea, 1-butyl-3-(3,4-dichlorophenyl)-1-methyl-
N-Serve	2-chloro-6-(trichloromethyl) pyridine
Paraquat	Bipyridinium compounds, 1,1-dimethyl-4,4'—— ... dimethyl sulfate or dichloride
Parathion	Phosphorothioic acid, O,O-diethyl O-p-nitrophenyl ester
PCNB	Benzene, pentachloronitro-
Phaltan	Phthalimide, N-[(trichloromethyl)thio]-
Phorate	Phosphorodithioic acid, O,O-diethyl S-[(ethylthio)methyl] ester
Ronnel	Phosphorothioic acid, O,O-dimethyl O-2,4,5-trichlorophenyl ester
Schradan	Pyrophosphoramide, octamethyl-
Silvex	Propionic acid, 2-(2,4,5-trichlorophenoxy)-
Simazine	s-Triazine, 2-chloro-4,6-bis(ethylamino)-
2,4,5-T	Acetic acid, 2,4,5-trichlorophenoxy-
TCA	Acetic acid, trichloro-
TDE	Ethane, 1,1-dichloro-2,2-bis(p-chlorophenyl)-
TEPP	Ethyl pyrophosphates
Thiodan	5-Norbornene-2,3-dimethanol, 1,4,5,6,7,7-hexachloro-, cyclic sulfite
Tillam	Carbamic acid, butylethylthio-S-propyl ester

Table XVI. (continued)

Pesticide	Chemical designation
Tordon	4-Amino-3,5,6-trichloropicolinic acid
Toxaphene	Chlorinated camphene containing 67 to 69 percent chlorine
Trichlorfon	Phosphonic acid, (2,2,2-trichloro-1-hydroxyethyl)-, dimethyl ester
Trifluran	Trifluoro-2,6-dinitro-N,N,-dipropyl-p-toluidine
Vernam	Carbamic acid, dipropylthio-, S-propyl ester
Zytron	Phosphoroamidothioic acid, isopropyl-, O-(2,4-dichloro-phenyl)-, O-methyl ester

Summary

Pesticides do not always interact with aquatic microorganisms as predicted. Generally all pesticides are toxic to all microorganisms at some dosage, the adage "the poison is in the dosage" holding true. Toxicity as measured in the several reported studies includes changes in growth rate, metabolic rate, and photosynthesis.

The phenylureas are the most toxic herbicides to phytoplankton, while surprisingly the cyclodienes are the most toxic insecticides. DDT can reduce photosynthesis in phytoplankton, and is also the most toxic material to many crustaceae.

Aquatic microorganisms absorb and concentrate pesticides from water apparently inversely related to the water solubility of the compound, DDT being the notable object of numerous studies. Living organisms do not seem to be any more efficient than dead organisms in this seemingly nonspecific, physical property of microorganisms.

Metabolism of pesticides in microorganisms is as varied as in vertebrates. Again it appears that any of the small forms will metabolize any of the pesticides to some extent, perhaps with the exception of dieldrin. In the case of DDT there are probably two routes of metabolism: aerobic, leading to the formation of DDE, and anaerobic, which produces TDE. In the case of actinomycetes metabolism occurred only during the active growth phase stopping completely when growth ceased. Phosphate insecticides are readily metabolized by all bacteria, actinomycetes, and fungi and algae examined. The five classes of herbicides are probably attacked by all forms, some being highly selective.

References

AHMED, M. K., and J. E. CASIDA: Metabolism of some organophosphate insecticides by microorganisms. J. Econ. Entomol. 51, 59 (1958).

ALEXANDER, M.: Microbiology of pesticides and related hydrocarbons. In: Principles and applications in aquatic microbiology. Proc. Rudolfs Research Conf. Rutgers, N.J. New York: Wiley (1964).

ANONYMOUS: Pesticide-wildlife studies. *U.S. Department of Interior*, Fish and Wildlife Service, Circ. 167 (1963).

AUDUS, L. J., Ed.: The physiology and biochemistry of herbicides. London and New York: Academic Press (1964).

BUTLER, P. A.: Effects of herbicides on estuarine fauna. S. Weed Control Conf. Proc. 18, 576 (1965a).

—— Commercial fishery investigations. *U.S. Department of Interior*, Fish and Wildlife Service, Circ. 226, p. 65 (1965 b).

—— Pesticides in the estuary. Proc. Marsh. Estuary Mgt. Symp., p. 120 (1967).

——, and P. F. SPRINGER: Pesticides—A new factor in coastal environments. Trans. 28th N. Amer. Wildlife Conf., p. 378 (1963).

CABEJSZEK, I., and J. STANISLAWSKA: Effect of methyl parathion (*p*-nitrophenyl *O,O*-dimethyl thionophosphate) on water-borne organisms. Roczniki Panstwowego Zakladu Higieny 17, 353 (1966).

—— Effects of thometon (*O,O*-dimethylthio-phosphate 2-ethyl mercaptoethyl) on water organisms. Roczniki Panstwowego Zakladu Higieny 18, 155 (1967).

CHACKO, C. I., and J. L. LOCKWOOD: Accumulation of DDT and dieldrin by microorganisms. Can. J. Microbiol. 13, 1123 (1967).

—— ——, and M. ZABIK: Chlorinated hydrocarbon pesticides: Degradation by microbes. Science 154, 893 (1966).

COON, F. B.: Private communication (1969).

COPE, O. B.: Contamination of the freshwater ecosystem by pesticides. J. Applied Ecol. 3 (Suppl.), 33 (1966).

COWELL, B. C.: The effects of sodium arsenite and silvex on the plankton populations in farm ponds. Amer. Fish. Soc. Trans. 98, 371 (1965).

CRANCE, J. H.: The effects of copper sulfate on Microcystis and zooplankton in ponds. Progr. Fish-Culturist 25, 198 (1963).

DE MARCO, J., J. M. SYMONS, and G. G. ROBECK: Behavior of synthetic organics in stratified impoundments. Amer. Water Works Assoc. J. 59, 965 (1967).

EDWARDS, C. A.: Insecticide residues in soil. Residue Reviews 13, 83 (1966).

FROBISHER, M., JR.: Fundamentals of bacteriology, 4th ed. Philadelphia: W. B. Saunders Co. (1949).

FUNDERBURK, H. H., JR., and G. A. BOZARTH: Review of the metabolism and decomposition of Diquat and Paraquat. J. Agr. Food Chem. 15, 563 (1967).

GOTTLIEB, D.: The disappearance of antibiotics from soil. Abstr., Phytopathol. 42, 9 (1952).

GREGORY, W. W., JR., J. K. REED, and L. E. PRIESTER, JR.: Accumulation of parathion and DDT by some algae and protozoa. J. Protozool. 16, 69 (1969).

GUENZI, W. D., and W. E. BEARD: Anaerobic biodegradation of DDT to DDD in soil. Science 156, 1116 (1967).

GUNNER, H. B., and B. M. ZUCKERMAN: Degradation of 'Diazinon' by synergistic microbial action. Nature 217, 1183 (1968).

GUNTHER, F. A., W. E. WESTLAKE, and P. S. JAGLAN: Reported solubilities of 738 pesticide chemicals in water. Residue Reviews 20, 1 (1968).

HARDY, J. L.: Effect of tordon herbicides on aquatic chain organisms. Down to Earth 22, 11 (1966).

HICKEY, J. J., J. A. KEITH, and F. B. COON: An exploration of pesticides in a Lake Michigan ecosystem. J. Applied Ecol. 3 (Suppl.) 141 (1966).

HILL, D. W., and P. L. McCARTY: Anaerobic degradation of selected chlorinated hydrocarbon pesticides: J. Water Pollution Contr. Fed. 39, 1259 (1967).

183

HIRAKOSO, S., I. KITAGO, and C. HARINASUTA: Inactivation of insecticides by bacteria isolated from polluted waters where the mosquito larvae breed in large number. Med. J. Malaya 22, 249 (1968).

JONES, B. R., and J. B. MOYLE: Population of plankton animals and residual chlorinated hydrocarbons in soils of six Minnesota ponds treated for control of mosquito larvae. Trans. Amer. Fish. Soc. 92, 3 and 121 (1963).

KALLMAN, B. J., and A. K. ANDREWS: Reductive dechlorination of DDT to DDD by yeast. Science 141, 1050 (1963).

KASAHARA, S.: Studies on the biology of the parasitic c pepod, Lernaea cyrinacea Linnaeus, and the methods of controlling this parasite in fish culture. Contr. Fish. Lab., Faculty of Agr., Univ. of Tokyo 3, 103 (1962).

KAUFMAN, D. D.: Structure of pesticides and decomposition by soil microorganisms. In: Pesticides and their effects on soils and water. Amer. Soc. Agron. Special Publ. No. 8. Symposium papers sponsored by Soil Sci. Soc. Amer. (1966).

KEARNEY, P. C.: Metabolism of herbicides in soils. In: Organic pesticides in the environment. Adv. Chem. Series 60, 250 (1966).

KEIL, J. E., and L. E. PRIESTER: DDT uptake and metabolism by a marine diatom. Bull. Environ. Contamination Toxicol. 4, 169 (1969).

KO, W. H., and J. L. LOCKWOOD. Accumulation and concentration of chlorinated hydrocarbon pesticides by microorganisms in soil. Can. J. Microbiol. 14, 1075 (1968).

LAMANNA, C., and M. F. MALLETTE: Basic bacteriology, ref. p. 61. Baltimore: Williams and Wilkins (1965).

LAWRENCE, J. M.: Aquatic herbicide data. Agr. Handbook No. 231 (1962).

LAZAROFF, N.: Algal response to pesticide pollutants. Bacteriol. Proc. 48, C149 (1967).

LICHTENSTEIN, E. P., K. R. SCHULZ, T. W. FUHREMANN, and T. T. LIANG: Biological interaction between plasticizers and insecticides. J. Econ. Entomol. 62, 761 (1969).

—— ——, R. F. SKRENTNY, and Y. TSUKANO: Toxicity and fate of insecticide residues in water. Arch. Environ. Health 12, 199 (1966).

LUCZAK, J., and J. MALESZEWSKA: Effect of Thiometon (O,O-dimethylthiophosphate 2-ethyl mercaptoethyl) on physio-chemical properties and development of bacteria in water. Roczniki Panstwowego Zakladu Higieny 18, 151 (1967).

MACKIEWICZ, M., K. H. DEUBERT, H. B. GUNNER, and B. M. ZUCKERMAN: Study of parathion biodegradation using gnotobiotic techniques. J. Agr. Food Chem. 17, 129 (1969).

MARTIN, J. P.: Influence of pesticides on soil microbes and soil properties. In: Pesticides and their effects on soils and water. Amer. Soc. Agron. Special Publ. No. 8, p. 95. Symposium papers sponsored by Soil Sci. Soc. Amer. (1966).

MATSUMURA, F., G. M. BOUSH, and A. TAI: Breakdown of dieldrin in the soil by a microorganism. Nature 219, 965 (1968).

MACRAE, I. C., and M. ALEXANDER: Microbial degradation of selected herbicides in soil. J. Agr. Food Chem. 13, 72 (1965).

——, K. RAGHU, and E. M. BAUTISTA: Anaerobic degradation of the insecticide lindane by Clostridium sp. Nature 221, 859 (1969).

—— ——, and T. F. CASTRO: Persistence and biodegradation of four common isomers of benzene hexachloride in submerged soils. J. Agr. Food Chem. 15, 911 (1967).

MENDEL, J. L., and M. S. WALTON: Conversion of p,p'-DDT to p,p'-DDD by intestinal flora of the rat. Science 151, 1527 (1966).

MISKUS, R. P., D. P. BLAIR, and J. E. CASIDA: Conversion of DDT to DDD by bovine rumen fluid, lake water, and reduced porphyrins. J. Agr. Food Chem. 13, 481 (1965).

MULLIGAN, H. F.: Management of aquatic vascular plants and algae. Internat. Symp. on Eutrophication, Madison, Wis. (1967).

MUNNECKE, D. E.: Fungicides. In D. C. Torgeson (ed.), Vol. 1. New York: Academic Press (1966).

ODUM, W. E., G. M. WOODWELL, and C. F. WURSTER: DDT residues absorbed from organic detritus by fiddler crabs. Science 164, 576 (1969).

PIERCE, MADELENE: The effect of the weedicide Kuron upon the flora and fauna of two experimental areas of Long Pond, Dutchess County, N.Y. N.E. Weed Control Conf. Proc. 12, 338 (1958).

—— Progress report of the effect of Kuron upon the biota of Long Pond, Dutchess County, N.Y. N.E. Weed Control Conf. Proc. 14, 472 (1960).

PORTER, J. R.: Bacterial chemistry and physiology, p. 407. New York: Wiley (1946).

PRAMER, D.: The persistence and biologcial effects of antibiotics in soil. Applied Microbiol. 6, 221 (1958).

RAGHU, K., and I. C. MACRAE: Biodegradation of lindane in submerged soils. Science 154, 263 (1966).

—— —— The effect of the gamma-isomer of BHC upon the microflora of submerged rice soil. I. Effect upon algae. Can. J. Microbiol. 13, 173 (1967 a).

—— —— The effect of the gamma-isomer of BHC upon the microflora of submerged rice soil. II. Effect upon nitrogen mineralisation and fixation and selected bacteria. Can. J. Microbiol. 13, 625 (1967 b).

REICHEL, W. L., T. G. LAMONG, E. CROMARTIE, and L. N. LOCKE: Residues in two bald eagles suspected of pesticide poisoning. Bull. Environ. Contamination Toxicol. 4, 24 (1969).

REYNOLDS, L. M.: Polychlorobiphenyls (PCB's) and their interference with pesticide residue analysis. Bull. Environ. Contamination Toxicol. 4, 128 (1969).

RISEBROUGH, R. W., D. B. PEAKALL, S. G. GERMAN, M. N. KIRVEN, and P. REICHE: Polychlorinated biphenyls in the global system. Nature 220, 1098 (1968).

——, P. REICHE, and H. S. OLCOTT: Current progress in the determination of the polychlorinated biphenyls. Bull. Environ. Contamination Toxicol. 4, 192 (1969).

ROBERTS, J. E., R. D. CHISHOLM, and L. KOTLITSY: Persistence of insecticides in soil. J. Econ. Entomol. 55, 153 (1962).

ROBINSON, J., A. RICHARDSON, A. N. CRABTREE, J. C. COULSON, and G. R. POTTS: Organochlorine residues in marine organisms. Nature 214, 1307 (1967).

ROGOFF, M. H.: Oxidation of aromatic compounds by soil bacteria. Adv. Applied Microbiol. 3, 193 (1961).

SCHECTER, M. S.: Private communication (1969).

SCHWARTZ, H. G., JR.: Microbial degradation of pesticides in aqueous solutions. J. Water Pollution Control Fed. 39, 1701 (1967).

SEAMAN, D. E., and T. M. THOMAS: Absorption of herbicides by submersed aquatic plants. Proc. Calif. Weed Conf., p. 11 (1966).

SETHUNATHAN, N., and I. C. MACRAE: Some effects of diazinon on the microflora of submerged soils. Plant and Soil 30, 109 (1969).

SWEENEY, R. A.: Metabolism of lindane by unicellular algae. Proc. 12th Conf. Great Lakes Research (1968).

TATUM, W. M., and R. D. BLACKBURN: Preliminary study of the effects of diquat on the natural bottom fauna and plankton in two subtropical ponds. S.E. Assoc. Game & Fish Comm. Proc. Ann. Conf., p. 16 (1962).

THIEGS, B. J.: Microbial decomposition of herbicides. Down to Earth (Dow Chemical Co.), Fall issue, p. 7 (1962).

TOTH, S. J., and D. N. RIEMER: Precise chemical control of algae in ponds. J. Amer. Water Works Assoc. 60, 367 (1968).

UKELES, R.: Growth of pure cultures of marine phytoplankton in the presence of toxicants. Applied Microbiol. 10, 532 (1962).

WARE, G. W., M. K. DEE, and W. P. CAHILL: Water florae as indicators of irrigation water contamination by DDT. Bull. Environ. Contamination Toxicol. 3, 333 (1968).

WATSON, G. H., and W. B. BOLLEN: Effect of copper sulfate weed treatment on bacteria in lake bottoms. Ecology 33, 522 (1952).

WEDEMEYER, G.: Dechlorination of DDT by *Aerobacter aerogenes*. Science 152, 647 (1966).

WESTLAKE, W. E., and F. A. GUNTHER: Organic pesticides in the environment. Adv. Chem. Series 60, 110 (1966).

WOODFORD, E. K., and G. R. SAGAR (Eds.): Herbicides and the soil. Oxford: Blackwell Scientific (1960).

WOODWELL, G. M., C. F. WURSTER, JR., and P. A. ISAACSON: DDT residues in an east coast estuary: A case of biological concentration of a persistent insecticide. Science 156, 821 (1967).

WURSTER, C. F., JR.: DDT reduces photosynthesis by marine phytoplankton. Science 159, 1474 (1968).

YASUNO, M., S. HIRAKOSO, M. SASA, and M. UCHIDA: Inactivation of some organophosphorous insecticides by bacteria in polluted water. Japan J. Expt. Med. 35, 545 (1965).

ROBERT C. HARRISS
DAVID B. WHITE
ROBERT B. MACFARLANE

Mercury Compounds Reduce Photosynthesis by Plankton

Mercury pollution is presently a serious problem in many parts of the world. Humans have died as a result of eating fish from mercury-contaminated coastal areas of Japan, high concentrations of mercury in fish and birds have been traced to industrial and agricultural discharges in Scandinavia, and at least 17 states in the United States have banned fishing in contaminated waters or warned against eating fish and shellfish contaminated with mercury. The sources and environmental pathways of mercury in the affected areas of Japan and Scandinavia have been studied (1). Investigations of mercury pollution in the United States are beginning.

An almost total lack of information on the biologic effects of mercury prevents the establishment of adequate water-quality standards. The Bureau of Water Hygiene, United States Public Health Service, and the Soviet Union have tentatively adopted a standard of

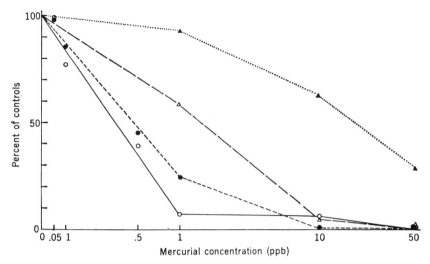

Fig. 1. Photosynthesis by *Nitzschia delicatissima*, a marine diatom, measured by uptake of carbon-14 relative to uptake by controls, after 24 hours of exposure to the following mercurials: Filled triangles, diphenylmercury; open triangles, phenylmercuric acetate; filled circles, methylmercury dicyandiamide; open circles, MEMMI.

5 parts per billion (ppb) for mercury in drinking water (2). In Japan a maximum allowable concentration of 10 ppb for methylmercury in industrial waste water has been adopted for the protection of humans (3). The present water-quality standards for mercury are inadequate to protect phytoplankton, organisms which are basic elements of almost all aquatic food chains.

Our experiments were designed to evaluate the acute effects of four commonly used organomercurial fungicides on a marine diatom, *Nitzschia delicatissima* Cleve, isolated from waters near Puerto Rico, and a naturally occurring phytoplankton population taken from Lake Jackson, a shallow freshwater lake near Tallahassee, Florida. The dominant genera in the freshwater phytoplankton population included *Merismopedia* sp. (Agmenellum), *Navicula* sp., *Crucigenia* sp., *Staurastrum* sp., and *Ankistrodesmus* sp. These phytoplankton were exposed to concentrations varying from 0 to 50 ppb of the following organomercurial compounds: phenylmercuric acetate (PMA or Phix); methylmercury dicyandiamide (Panogen); *N*-methylmercuric-1,2,3,6-tetrahydro-3,6-methano-3,4,5,6,7,7-hexachlorophthalimide (MEMMI); and diphenylmercury.

All experiments were conducted in a growth chamber equipped with a balanced bank of Gro-lux and Daylight (Sylvania) fluorescent lights. The light intensity was maintained at 1×10^4 erg cm^{-2} sec^{-1}. The chamber was programmed for a photoperiod of 12 hours light, 12 hours dark and maintained at $25° \pm 0.5°C$.

When the concentrations of *N. delicatissima* reached 7.5×10^4 cell/ml and the population was in the logarithmic phase of growth, 50-ml portions were exposed to varying concentrations of the fungicides. After 24 hours of exposure to the mercurials, 5 μc of [^{14}C]NaHCO$_3$ was added to each of two light bottles and to one dark bottle. The flasks were returned to the growth chamber for 5 hours of exposure to light, then the contents were filtered, and the radioactivity was counted on a Picker proportional counter.

The Lake Jackson sample was exposed to the four mercurials and grown in 500-ml flasks. At 24, 72, and 120 hours after initial exposure, 150 ml was withdrawn from each flask. A 50-ml portion was put into each of two light bottles and one dark bottle. Then 5 μc of [^{14}C]NaHCO$_3$ was added to each bottle. All bottles were then returned to the growth chamber for 5 hours of exposure to light, then the contents were filtered, and the radioactivity was counted on a Picker proportional counter.

In all experiments the count obtained from dark bottles and the background count were subtracted from the average of the two light bottles. The effect of the mercurials on photosynthesis and growth is calculated by dividing the net count at each concentration by the net count of the control sample and is expressed as uptake of carbon-14 as percent of controls.

Of the four mercurials, studied, diphenyl mercury was the least toxic (Figs. 1 and 2). At 1 ppb of the other three mercurials, a significant reduction in photosynthesis and growth is observed in cultures of *N. delicatissima* and the freshwater phytoplankton. At 50 ppb essentially all uptake of inorganic carbon is stopped; cell counts also indicate complete inhibition of growth. The reproducibility of these experiments, determined by exposing five identical samples from Lake Jackson to 1 part of methylmercury dicyandiamide per billion, was $\pm$ 11 percent. In related experiments, not shown in the figures, the toxicity of mercuric chloride to the Lake Jackson phytoplankton population was similar to that of diphenylmercury. We have also noted that the toxicity of any particular mercurial compound decreases with increasing cell concentration in lake samples in a man-

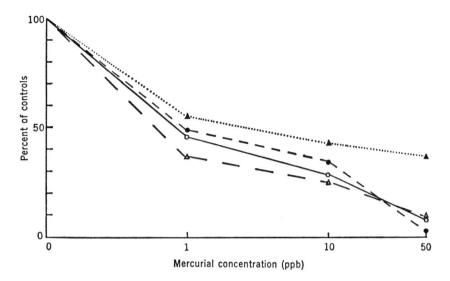

Fig. 2. Photosynthesis by a freshwater phytoplankton population, measured by uptake of carbon-14 relative to uptake by controls, after 120 hours of exposure to the following mercurials: Filled triangles, diphenylmercury; open triangles, phenylmercuric acetate; filled circles, methylmercury dicyandiamide; open circles, MEMMI.

ner similar to effects of chlorinated hydrocarbons (4).

Our studies indicate that at least some marine and freshwater phytoplankton species are sensitive to much lower concentrations of mercurial compounds than have been observed to affect fish in short-term tests of toxicity (5). It is also clear that concentrations of mercurial compounds well below the proposed water-quality standards can have detrimental effects on phytoplankton. We conclude that the use of organomercurial compounds in any way that permits their discharge into natural waters should be stopped as soon as possible. The long-term effects of mercury pollution below concentration of 1 ppb must be determined to establish adequate water-quality standards.

References and Notes

1. A. G. Johnels and T. Westermark, in *Chemical Fallout*, M. W. Miller and G. Berg, Ed. (Thomas, Springfield, Ill., 1969), pp. 221–244.
2. Anonymous, *J. Amer. Water Wks. Assoc.*, 1970, p. 285.
3. Anonymous, *Water Newslett.*, 20 August 1969, p. 1.
4. C. F. Wurster, *Science* 159, 1474 (1968).
5. A. G. Johnels, T. Westermark, W. Berg, P. I. Person, B. Sjöstrand, *Oikos* 18, 323 (1967).
6. We thank W. Glooschenko for assistance in phytoplankton identifications and for supplying the diatom culture and J. W. Winchester for stressing the importance of the results to water-quality standards.

Contemporaneous Disequilibrium, a New Hypothesis to Explain the "Paradox of the Plankton"*

Peter Richerson, Richard Armstrong   and Charles R. Goldman

The structure and functional relationships of lake phytoplankton exhibit many puzzling phenomena. One of the most perplexing is the unexpectedly high diversity encountered in even small samples of phytoplankton. G. E. Hutchinson[1] has dealt at length with this problem terming it "the paradox of the plankton." Briefly stated, this paradox is that the examination of a small volume of water (e.g., 10 ml) usually yields a list of some tens of species where the competitive exclusion principle[2] might lead us to expect only one or a few species. One can argue that the epilimnion of a lake is as nearly homogenous as any habitat can be expected to be due to turbulent mixing and that the competitive exclusion principle would, if its postulates were met, lead to the exclusive occupation of the habitat by a single species best adapted for living there.

The competitive exclusion principle has two postulates that may explain its apparent inapplicability to the phytoplankton. It assumes that the competing species are at equilibrium and that the axiom of inequality holds (that is, that two material systems are never exactly equivalent and therefore cannot have a competition coefficient of zero). Hutchinson believes that the main answer to the paradox of the plankton lies in the violation of the first assumption. Since conditions change quite rapidly in the plankton habitat, perhaps one, and then another, organism is the superior competitor, but in such rapid succession that no one organism has the advantage long enough to cause the extinction of the others. No evidence seems to contradict this explanation, although decisive observational support is also lacking.

Riley[3] offers another explanation centering around the second assumption. He believes that natural selection has caused phytoplankton to approach asymptotically some upper limit of efficiency which makes differences between species so small that extinction, even in an equilibrium state, would proceed at a very slow rate. Hutchinson[4] also discusses two additional mechanisms—symbiosis and niche diversification—and believes that it is possible that some phytoplankton are meroplanktonic and are not able to reproduce in the plankton indefinitely. There is some reason to think that many phytoplankton are just opportunistic forms that happen to find the pelagic habitat suitable for growth for restricted periods.

The various hypotheses are not mutually exclusive and any one can help to explain "the paradox of the plankton." Available evidence indicates that all hypotheses are plausible; such evidence is insufficient to assign relative importance to any hypothesis, much less eliminate any entirely. Hutchinson notes that this unexpected diversity "perhaps has never been fully explained." This view is clearly not an overstatement of the difficulty encountered.

Our data[5] from Castle Lake, a mesotropic subalpine lake of 0.2 km² in Northern California, combined with the previous evidence and certain observations made by the authors at Lake Tahoe suggest a contemporaneous disequilibrium model similar to Hutchinson's. 72 replicate samples, distributed from six different epilimnetic stations, were taken. Two indices of patchiness, Fisher's k statistic and the variance to mean ratio, were calculated for these data, as shown in Table 1. Fisher's k was calculated by a maximum likelihood method[6] with a digital computer. Many of the species found in such samples were very patchily distributed. The pattern produced must be attributed to reproduction as most of the species involved were diatoms and other nonmotile or slightly motile forms. Although the subject of patchiness in phytoplankton has had much less attention than in zooplankton, some other examples of phytoplankton superdispersion are available.[7] The existence of patchiness is contrary to Hutchinson's implicit assumption about phytoplankton. He states the problem as one in which many species must be maintained in the face of a "relatively isotropic or unstructured environment all competing for the same sorts of materials." If patches based upon reproductive patterns are possible, there must be enough structural stability, relative to the reproductive rate of phytoplankton in the environment, for such patches to be established. There is some tendency for the rarer organisms to be either very patchy or random to slightly underdispersed, suggesting that such temporary niches are most important in maintaining these rarer organisms.

The potential doubling times of algae are very short. In Lake Tahoe, an ultraoligotrophic lake, average turnover rates for the whole water column (determined from the ratio of ¹⁴C productivity to carbon biomass, assuming carbon to be 13% of the total biomass) are as low as 1.4 days; individual samples are

TABLE 1

Species	Mean	Fisher's k^{10}	Variance to mean
Botryoccus braunii	34.89	7.70	5.35
Cosmarium bioculatum	0.56	*	0.84
Cosmarium circulare	46.93	36.26	2.29
Oocystis lacustris	0.17	—	1.00
Oocystis naeglii†	2.65	1.79	2.38
Quadrigula chodati†	4.51	0.29	12.04
Quadrigula sp.†	0.21	0.17	4.93
Staurastrum curvatum	0.38	—	1.00
Staurastrum brevispinum†	0.19	1.90	1.09
Dinobryon sertularia†	45.57	1.54	23.46
Achnanthes minutissima†	0.81	1.17	1.71
Achnanthes linearis	1.11	—	0.96
Asterionella formosa†	0.08	0.02	4.25
Cyclotella meneghiniana†	0.78	2.22	1.51
Cymbella gracilis	0.08	—	0.92
Cymbella turgida	0.08	—	0.92
Cymbella ventricosa†	0.15	0.56	1.21
Diatoma anceps†	0.42	3.41	1.12
Diatomella balfouriana†	0.08	0.08	1.58
Epithemia zebra†	0.17	0.79	1.17
Eunotia incisa†	0.08	0.26	1.25
Fragilaria brevistriata†	0.53	1.41	1.26
Fragilaria construens†	0.31	0.03	5.24
Fragilaria virescens†	0.08	0.02	2.92
Fragilaria pinnata†	0.53	0.97	1.58
Gomphonema ventricosum	0.25	—	0.86
Navicula capitata	0.14	—	0.86
Navicula cryptocephala	1.54	—	0.68
Navicula lanceolata	1.40	9.89	1.14
Navicula minima	0.29	—	0.90
Navicula monmouth-stodderi†	0.10	0.42	1.19
Navicula radiosa	0.50	—	1.00
Pinnularia biceps†	0.08	0.08	1.58
Surirella intermedia	0.08	—	0.91
Synedra radians b.†	0.47	4.07	1.12
Synedra radians c.†	0.50	9.00	1.06
Tabellaria flocculosa†	0.42	0.14	5.12
Nitzschia amphibia	0.69	10.59	1.07
Opephora martyi†	1.72	0.04	79.89
Ceratium hirundinella	0.61	—	0.93
Anabaena affinis†	12.11	0.39	17.25
Chroococcus limneticus distans	10.22	5.22	2.92
Gloeocapsa granosa†	1.97	1.13	3.90
Merismopedia glauca†	2.54	0.07	16.06
Polycystis aeruginosa	516.03	68.91	8.39
Euglena sp.	61.67	5.00	12.35

Means and indices of patchiness for algae in Castle Lake. Portion of slide counted is equivalent to 3 ml of lake water and the total number of samples counted was 72. Organisms with less than 6 total occurrences omitted.

* k was not calculated for species with variance to mean ratios less than 1.

† k of less than 5, indicating appreciable departure from randomness.

occasionally observed with turnover rates as low as 8 hr. Phytoplankton diversity in Tahoe appears uncorrelated with zooplankton diversity but related to grazing pressure, since it is highest during the summer–fall zooplankton bloom, especially during high populations of *Daphnia pulicaria*. Fig. 1 shows the high,

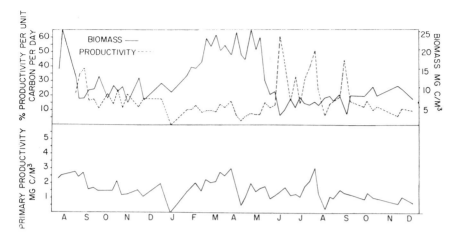

FIG. 1. Primary productivity, productivity per unit biomass, and biomass curves for phytoplankton in Lake Tahoe.

but extremely variable, productivity/biomass ratio (turnover rate) during the periods of high zooplankton populations throughout the water column from May until December. Fig. 2 is a typical productivity/unit biomass curve for Lake Tahoe showing irregular layers of high turnover. Other studies[8] show a high degree of variability in the vertical structure of water masses, corresponding to observations of divers who report striking layering effects of the plankton. Thus, there is at least as much evidence to support the hypothesis that the major diversifying factor is contemporaneous heterogeneity, random or chaotic perhaps, but with enough persistence to allow many species to exploit the whole habitat simultaneously. The observed relation of phytoplankton diversity to zooplankton biomass (grazing pressure) and turnover rate is especially interesting because it suggests that decreasing the doubling time of the algal biomass enables particular species to exploit smaller, or less persistent, uniquely favorable water masses through more rapid reproduction. This relationship is also interesting since it is the opposite of the one predicted by Margalef,[9] who finds that exploitation generally reduces diversity. The concepts of within- and between-habitat diversity[10] are useful in this case. Grazing may indeed make the diversity in one small unit of water less according to Margalef's theory, thus reducing within-habitat diversity, but it may increase

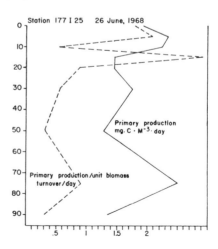

FIG. 2. Productivity, and productivity per unit biomass, curves for Lake Tahoe. Depth is in meters.

195

measured diversity by reducing the size of habitats, thus increasing between-habitat diversity. Since within-habitat diversity is probably small in the plankton, the result is increased diversity. The universally observed patchiness of zooplankton must also be one of the many factors imparting uniqueness to these water masses.

This hypothesis differs from Hutchinson's by stressing the contemporaneous, rather than temporal, heterogeneity of the plankton habitat. The epilimnion of a lake is probably not completely homogenous on a time scale of a few hours, but rather provides a number of unique niches. Such a habitat is, however, quite unstable, obliterating these niches and reconstituting them at frequent, random, intervals.

It must be stressed that the two nonequilibrium hypotheses are not contradictory, but rather reinforce one another. The patch of water in which a particular species blooms and approaches a monospecific equilibrium has both spatial and temporal dimensions. It must be large enough in both its spatial and temporal dimensions relative to the reproductive rate of the species concerned and the rate of turbulent transport for one, or a very few, species to reproduce disproportionately with respect to the others present.

* P. R. was supported by a NSF predoctoral fellowship. The field work was also supported by a Federal Water Pollution Control Agency DI grant 16010 DBU and NSF grant GB6422X to C. R. Goldman.

[1] Hutchinson, G. E., *Amer. Natur.*, **95**, 137 (1951).

[2] Hardin, G., *Science*, **131**, 1292 (1960).

[3] Riley, G. A., "Marine Biology I." *Proceeding of the First International Interdisciplinary Conference*, ed. G. A. Riley, (American Institute Biological Sciences, 1963).

[4] Hutchinson, G. E., *A Treatise on Limnology* (New York, John Wiley & Sons, 1967), vol. 2.

[5] Armstrong, R., Ph.D. dissertation, University of California, Davis (1969).

[6] Bliss, C. I., and R. A. Fisher, *Biometrics*, **9**, 177 (1953).

[7] Lund, J. W. G., C. Kipling, and E. D. LeCren, *Hydrobiologia*, **11**, 147 (1958); Margalef, R., *Perspectives in Marine Biology*, ed. A. A. Buzzati-Traverso (University of California Press, 1960), pp. 323–349.

[8] Lovett, J. R., *Limnol. Oceanogr.*, **13**, 127 (1968); Baker, A. L., *Limnol. Oceanogr.*, **15**, 159 (1970).

[9] Margalef, R., *Perspectives in Ecological Theory* (Chicago: University of Chicago Press, 1968).

[10] MacArthur, R. H., *Biol. Rev.*, **40**, 510 (1965).

AUTHOR INDEX

Anderson, Judith, 151
Antia, N. J., 36
Armstrong, Richard, 192

Bailey-Watts, A. E., 83
Belcher, J. H., 83
Benson, A. A., 64
Berger, Wolfgang H., 93
Bindloss, M. E., 83

Carlucci, A. F., 42
Chandler, P. T., 64
Cherry, R. D., 53
Clarke, Thomas A., 7
Corner, E. D. S., 98
Cowey, C. B., 98
Cox, James L., 142

Davenport, Demorest, 68

Farkas, Tibor, 25
Flechsig, Arthur O., 7
Forward, Richard, 78
Fudge, H., 51
Fuller, G., 64

Goldman, Charles R., 192
Grigg, Richard W., 7

Harriss, Robert C., 187
Hecht, Alan D., 146
Herodek, Sandor, 25
Horiguchi, M., 132

Kalan, E. B., 64
Kalmakoff, J., 36
Kalnina, Z., 88
Kittredge, J. S., 132

Loeblich, A. R., III, 64

Macfarlane, Robert B., 187
Menzel, David W., 151

Orren, M. J., 53

Patrick, Ruth, 75
Patton, Stuart, 64
Polikarpov, G., 88

Randtke, Ann, 151
Richerson, Peter, 192
Roan, Clifford C., 157
Robinson, G. A., 71
Ryther, John H., 155

Savin, Samuel M., 146
Shannon, L. V., 53
Silbernagel, S. B., 42
Soutar, Andrew, 93

Ware, George W., 157
Watt, A., 36
White, David B., 187
Williams, P. M., 132
Wurster, Charles F., Jr., 137

KEY-WORD TITLE INDEX

Aminophosphonic Acids, Biosynthesis of, 132

Bioassay of Seawater, 42
Blue-Green Alga, Freshwater Primary Production by, 83

Chlorinated Hydrocarbons, Responses to, 151
Contemporaneous Disequilibrium, 192

DDT and Photosynthesis, 137
DDT Residue, 142
Diatom Community, Structure, 75
Dinoflagellate Behavioral Response, 68

Enolase Activity in Algae, 36

Fatty Acid Composition of Crustaceans, 25
Foraminifera, Oxygen-18 Studies, 146
Foraminifera, Production Rate Experiment, 93

Gonyaulax polyedra Food Value, 64
Gonyaulax tamarensis Lebour Distribution, 71

Lead-210 Presence, 53

Mercury Compounds and Photosynthesis, 187

Oxygen Supply Threat, 155

"Paradox of the Plankton," 192
Pesticide Interaction, 157
Polonium-210 Presence, 53
Project Sealab II, 7

Strontium-90 Concentration Factors, 88

Vitamin B12 in Seawater, Determining, 42

Zooplankton, Biochemical Analysis, 51
Zooplankton Production, Biochemical Study, 108